From The Stephenson Locomotive Society A

Eclectic Electrics

RAS Hennessey

First published in 2017

The Stephenson Locomotive Society

Bristol, England.

www.stephensonloco.org.uk

© The Stephenson Locomotive Society, 2017.

All rights reserved. No part of this publication may be reproduced or transmitted in any form or by any means, electronic or mechanical, including photocopying, recording, or any information retrieval system, without prior permission in writing from the publishers. Within the UK, exceptions are allowed in respect of any fair dealing for the purpose of research or private study, or criticism or review as permitted under the Copyright, Designs and Patents Act 1988.

The book is sold subject to the condition that it shall not, by way of trade or otherwise, be lent, re-sold, hired out or otherwise circulated without the publishers' prior consent in any form of binding or cover other than that in which it is published and without a similar condition including this condition being imposed on the subsequent purchaser.

Every effort has been made to trace or contact all copyright holders. The publishers will be pleased to make good any omissions or rectify any mistakes brought to their attention at the earliest opportunity.

Typesetting, design and production by John New.
www.island-publishing.co.uk

Printed by Print Team (Dorset) Ltd.
Unit 21, Mereside, Portland, Dorset, DT5 1PY, UK.

ISBN 978-0-903881-08-1

FOREWORD

Where would our world be without electricity? A similar question applies to the modern railway - where would it be without electric traction? The industry could not survive without electric power.

Moreover, the pedigree of the modern electric powered railway goes back a long way: Davidson's pioneering trials of 1842 in Scotland, took place only 13 years after *Rocket* had won the famous competition at Rainhill - a success that in turn ushered in the inter-city railway era!

The book is a compilation, introducing examples of the pioneering electrification schemes together with oddities of power supply. As the title suggests, an eclectic mix reflecting the range and diversity of interests covered by The Stephenson Locomotive Society and its members. The Society was born as the electrification era began to flourish; one of the first outings by members was to view the new London, Brighton & South Coast Railway (LBSCR) electric car sheds near Norwood; electric traction in all variants and flavours is therefore a fitting subject for a Society publication.

This second book in the series of extracts from material held within the Society archives, or written by members, is Part 1 of the collection regarding electrification matters. The author has written widely on the subject of electrification in particular, and variations from railway orthodoxy in general.

John R New. Production Editor. November 2017

CONTENTS

Foreword	ii
Original EMUs	3
The St Clair Tunnel Electrics	7
The General Electric Gearless Project	11
The Oerlikon Rod	18
Kinlochleven - A Pioneer Hydro-Electric Line And Its Setting	22
Three-Phase Ac Traction In The Americas	29
An Odd Voltage - The Rise and Fall of 2.4kV dc.	40
Henry Ford's Do-Do - a Unique and Extinct Species.	45
The Baltic Electrics	50
Very High Voltage Railways	59

IMAGES

All drawings and illustrations are from the Author's Collection unless individually credited.

COVER IMAGE:

One of the class EP-2 'gearless' electric locomotives threads its way through the Cascade Mountains with a transcontinental train running under the 3kV dc conductors of the 'Milwaukee Road' (Chicago Milwaukee St Paul & Pacific RR).

The image has been prepared from a colourised postcard; details of the original artist and publisher are unknown, although a date for the image is probably the early 1920s. All is long gone; the electrics, the railroad, and its western extension.

For further information regarding these locomotives, and an additional photograph, see 'The General Electric Gearless Project' chapter on pages 14-15.

For correspondence regarding this work please contact the *Journal* Editor using the email address journal@stephensonloco.org.uk or the SLS General Secretary, 23 Passmore Way, Tovil, Maidstone, KENT ME15 6 DZ.

This volume is published by **The Stephenson Locomotive Society**. For academic referencing needs our formal base is Bristol; however, the Society Registered Office address should not be used for routine postal communications regarding this volume.

ORIGINAL EMUS

Nothing to do with palaeontology or *Dromaius Novaehollandiae*, the soft-feathered flightless ones of Australia but, of course, the 'electric multiple unit' or, in railway argot, the EMU. This brief paper examines the origin of the EMU concept and when it first came to be adopted in the UK.

The invention of the EMU is generally accredited to Frank Sprague (1857-1934) a prolific, talented and successful American inventor. This accreditation is fair since Sprague worked hard and carefully on the concept which he had arrived at independently. Moreover, it was his eminently practical form of EMU that was adopted by the transportation world from 1897 onwards.

Even so, it is said that few things are new under the sun, which is certainly the case with the EMU. As early as 1883, possibly late 1882, Professor W E Ayrton addressed the Royal Institution in London in these terms, from a summary reported in *Nature* 11 January 1883,

> '… the employment of electricity … will enable a train to be driven with every pair of wheels, just as the employment of compressed air enables every pair of wheels to brake the train'.

In his early years, Sprague had been an officer in the US Navy and had been seconded for a while to the UK. He travelled often on the steam-laden 'Inner Circle' in London, coming to the conclusion that electric traction would be a definite improvement for this engineering marvel. It is said that his profound interest in electrification, in various forms, dated from that time. He appears to have first thought about the multiple unit system, as such, in late 1885, when he spoke to the Society of Arts in Boston on the subject.[1]

Before he hit upon a practical form of the multiple-unit concept, Sprague had built up a considerable stock of experience and inventions. These included improved electric lifts, the nose-suspended motor for trams and trains, the trolley wheel adorning spring-loaded booms on trams, the compound-wound dc motor, series-parallel control of dc motors, and early forms of regeneration. This is quite a 'bag' in its own right, and clearly the basis of many necessary features in the growth of electric traction, and the cities and suburbs dependent upon it.

Sprague's system was first adopted by Chicago's South Side 'L' (= Elevated Railway, 'El' in New York-ese) in 1897 where it was a resounding success. Sprague, who at first referred to his new system as 'distributed motive power', listed its many advantages, for example that a train composed of anything from one coach to many could be controlled from either end. Providing the train was composed of similar 'units' its acceleration and deceleration would be the same regardless of the number of carriages composing it.

The EMU promised 'economies of operation', for example in reducing crew, and therefore wage costs, although the Chicago Elevated Railway continued to employ assistant motormen for a few years. By dispensing with locomotives and running lighter trains there was less wear and tear to infrastructure, rails, ways and works. Sprague claimed that EMUs were simple to drive, 'a child of ten can handle full-sized trains' he asserted, and it was true enough as proved by his own young son.

The confident Sprague recommended to the management of the South Side L, 'Without further argument, I recommend on your road the absolute abolishment of the locomotive system … ' going on to say of his own system, 'the individual equipment of cars with one or more motors, so controlled one or any number of cars without regard to sequence or position can be coupled together indiscriminately, and from the leading end of any car the train system can be operated'.

The new system caught on quickly, especially in the land of its birth. The South Side L was soon emulated by New York's Brooklyn Elevated Rail Road and, back in Chicago the Metropolitan and Lake Street Ls. By 1904 the multiple unit system had been adopted by some 70 railways, operating 3,000 multiple control units between them.

Perhaps predictably, General Electric and its great rival Westinghouse soon produced their own versions of the system rivalling that of the Sprague Electric Company. Patent rights struggles were a characteristic feature of the age, a period of astonishing innovations. Eventually, General Electric bought the Sprague Electric Company for $2.5 million, and accommodation was made with Westinghouse.

British railway entrepreneurs were quick off the mark in adopting the EMU system. One good reason for this was the strongly dominant position of American capital and technology at the time. Both contenders for the UK's 'first EMU' system were backed by American investment, and were dependent on American electrical technology, the Central London Railway (CLR), and the Mersey Railway.

However, establishing the UK's 'first' EMU is not straightforward. Credit is generally accorded to the CLR which converted to EMUs from its original, none too successful, locomotive operation. The CLR was able to draw on the best advice since its electrical consultant was Horace Parshall, an American who had previously worked closely with Sprague. Parshall designed the CLR's original locomotives and later became Chairman of the company.[2]

The CLR opened in 1900 with a fleet of 28 gearless locomotives[3] and 150 trailer carriages, soon augmented by 18 more. After it had been found that its electric locomotives caused unacceptable vibration, the CLR tried out geared, axle-hung

An original Central London Railway (CLR) EMU manoeuvring at Wood Lane depot c1904. The driving compartment and cabin for switchgear and resistances appeared rather awkwardly 'tacked on' to standard CLR cars, livery LNWR-ish. The overhead conductors were for electric shunting locomotives, the CLR conductor was a third rail, centrally placed between running rails. In the distance, on the far left, one of the CLR's two oil-fired Hunslet 0-6-0Ts, built to tube gauge and fitted with condensing apparatus for tunnel work.

motors, slightly more successfully, and also two experimental EMUs in 1901. An official enquiry from the Board of Trade (the 'Vibration Committee') pronounced in favour of the EMU system (Also see p12). The CLR was entirely co-operative with public authorities during the vibration crisis, and quickly invested in the new EMU system.

Sixty-four new motor coaches ('motor cars' in CLR, American-influenced usage); 24 from Brown, Marshall who were soon to fade into the Metropolitan Amalgamated Railway Carriage & Wagon combine, and 40 from Birmingham Railway Carriage & Wagon Company, with later additions and adjustments.

The new stock started operations in April 1903; the last locomotive hauled trains went in June of that year. One immediate result of EMU working was obviating the need to uncouple and couple locomotives, even to run them round trains. At a stroke, the capacity of the line was enhanced by the flexible new operating system. Further flexibility came from the ability to halve six-car trains in off-peak periods, driving them from the motor-car, or a 'control trailer' – about half the CLR carriage stock was to be of this kind.

Another kind of flexibility made possible by the CLR EMUs was the ability of these sinuous, all-carriage trains to negotiate the extremely tight loop that took trains from Shepherds Bush to the Wood Lane terminus, and back again. Part of this layout was the 'Caxton Curve', still the tightest on the Underground, with a radius of 200ft.

En passant, the CLR took the bold step of installing electrically powered lifts in lieu of hydraulic ones. The contract for this lift installation went to Frank Sprague although he had sold out to Otis Elevator before installation started.[4]

Merseyside. The other contender for 'first (UK) EMU' is the Mersey Railway. Newly electrified by Westinghouse interests, it started operations with EMUs in May 1903. Thus, the Mersey Railway was the first in the UK (and Europe) to run all services with EMUs from the start, just ahead of the CLR reaching that condition, mid-June 1903. Perhaps honours ought to be divided equally.

The original Mersey Railway stock comprised 57 cars constructed by G F Milnes & Co, Castle Car Works, Hadley, Shropshire. The English connection really ended there: the cars were typical American 'subway' stock with clerestories, Baldwin bogies, Westinghouse air brakes, and (for the motor cars)

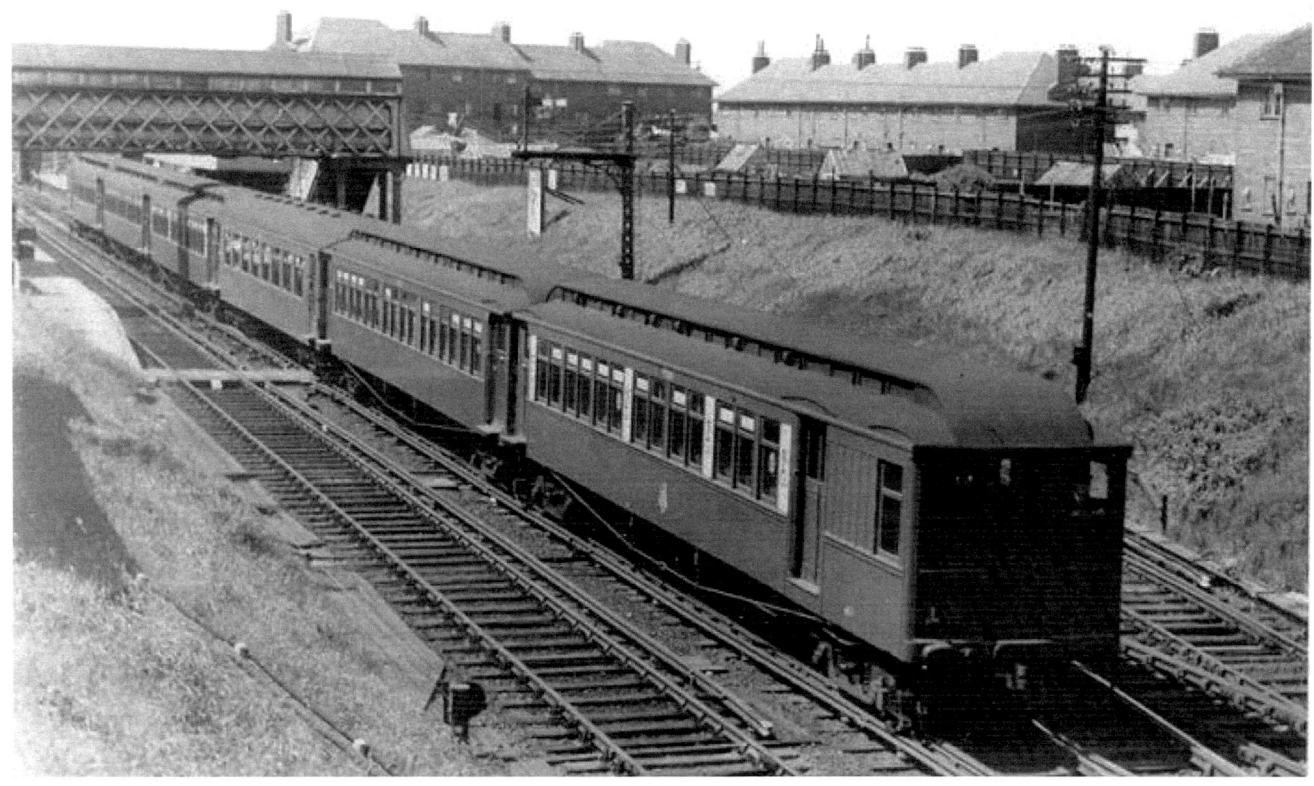

Mersey Railway EMU stock in BR livery on a New Brighton-Liverpool train leaving Birkenhead North, June 1959. Composed mostly of 1903 Milnes-British Westinghouse stock, the third coach, however, appears to be one of the 1923 Cravens which could be used interchangeably with the older EMUs. Photo: *John H Meredith*

two Westinghouse No.83 motors on each bogie; the multiple unit controls were also electro-pneumatic as opposed to the Sprague-General Electric approach which was electro-magnetic. In true multiple unit style, no power lines were carried along the train; each motor car picked up and used its own current from the conductor rail.

Each motor car mounted two large controller drums geared together, one for each bogie. The small 'master controller' employed by the driver sent low-voltage instructions to these large drums (they weighed one ton together) moving them on through ten steps of series-parallel control by means of an air-operated piston.

Given the crucial importance of compressed air in these pioneer EMUs (for brakes, sanding, and control) it may come as a surprise to learn that the trains did not compress air for themselves, but were replenished with pumped air at Liverpool Central, Rock Ferry, and Birkenhead Park, where storage reservoirs on the train were charged with air at 135lb sq in, lowered by reducing valves for usage on the train.

There is a third contender for the UK's 'first EMU', namely the Metropolitan District Railway which was trying out an experimental EMU, using Mersey-type equipment, on the Ealing & South Harrow Railway at about the same time, but the laurels of being the first line to go exclusively for EMU working in the UK, and for that matter in Europe, will forever adorn the name and reputation of the Mersey Railway.

Whether one classes the original Liverpool Overhead and Lancashire & Yorkshire Railway's Southport electric trains as 'multiple unit' is a matter of rather fine definition as both ran power cables along the train, controlling all motors from the same master controller. However, perhaps 'hybrid EMU' fits the bill; some of the modern TGV fleet has similar arrangements. For this reason, the North Eastern Railway's (NER) Tyneside electrification of 1904 might lay claim to being the first by a British main line company to employ the 'true Sprague' system; the NER's contractors being British Thomson-Houston, an arm of American General Electric.

By 1918 the EMU had come to stay in the UK; examples were operating on Tyneside (NER), around Liverpool and Manchester (Mersey, Liverpool Overhead Railway, Lancashire & Yorkshire Railway), to the North (Lancaster-Morecambe-Heysham) and, above all the London region – the Underground, Metropolitan, London, Brighton & South Coast Railway, London & South Western Railway, London & North Western Railway: however, these enterprises were all urban or suburban schemes at best. The practical long-distance EMU lay in the future, although speculation about it had already started, for example from the pen of J F Gairns in *The Railway Magazine*, 1919, who suggested, amongst others, EMU working for London-Birmingham-Liverpool-Manchester and for Glasgow-Edinburgh.

The 'EMU flyer' first appeared in Southern England, for example 'Brighton Belle' sets (5-BEL) and the London-Portsmouth 'Nelsons' (4-COR). The

At the cutting edge: part of the Japanese Railways East fleet of *Shinkansen* ('New Trunk Line') EMUs on parade, November 2007.

Southern Railway also invested in a wide range of suburban and middle-distance EMUs, some as new stock, others as older steam stock 'retreads'. A full list, compiled by Bruce Nathan, can be found in the SLS *Journal* No.853, September-October 2008, 'Southern Electric Classification'.

The EMU has come to stay 'big time' by the early-twenty-first century. Even so, it is a fairly venerable railway phenomenon. Because it self-propels, with one or more axles or coaches electrically powered, it is a form of 'locomotive', although far removed from a free-standing steam engine in form.

Taking a fairly broad interpretation of 'EMU' and its hybrid forms, it could sensibly be asserted that the 'EMU' is now at the leading edge of passenger train technology, exemplified by the Shinkansen bullet trains in Japan, the TGVs and British Class 373.

An early Stephenson Locomotive Society members outing (1912) was to the London, Brighton & South Coast Railway sheds at Norwood, where the Brighton's 'Elevated Electric' EMU ac stock was housed and inspected. Those broad-minded railwayists were witnessing the dawn of a motive power revolution, doubtless reconciled by its being set firmly on Brighton Line soil, a railway dear to the hearts of SLS pioneers[5].

NOTES

1. See, *Frank Julian Sprague, Electrical Engineer & Inventor* by William S Middleton and William D Middleton III, University of Indiana Press, 2009.

2. Horace Field Parshall 1867-1932, settled in the UK. He was closely involved with electric tramways, including those in Glasgow, Paisley, Isle of Thanet, Bristol and Newport, Monmouthshire, and electric power companies in Lancashire and Yorkshire. Being an 'electrical' he has received little recognition in steam-dominated railway literature.

3. See SLS *Journal*, No. 858-60. (July/August & November/December, 2009)

4. It may be of passing interest, but J N Maskelyne, President of the SLS 1925-60, worked as an engineer for Waygood-Otis, lift manufacturers.

5. Although outside the scope of this book that interest ultimately led to the preservation of the LBSCR steam locomotive, 214 *Gladstone*, now on display at the NRM, York.

THE ST CLAIR TUNNEL ELECTRICS

Back in January/February 2008 the extraordinary 0-10-0Ts of the St Clair Tunnel Co were part of an SLS *Journal* article on 'Stretch Tanks' (pp.11-20). Their electrically powered successors were no less unusual and interesting – and they lasted for half a century.

But first things first, some basic facts and data. The first St Clair Tunnel was constructed 1888-1891, later closed and replaced by a new version, opened in 1994. The purpose of both tunnels has been identical: to link Sarnia in Ontario, Canada to Port Huron, Michigan, USA by means of a subaqueous bore under the St Clair River, the international boundary. This river, dividing the USA from Canada, joins Lake Huron with Lake St Clair (and thence, Lake Erie). Before the tunnel was constructed the river caused a serious traffic bottleneck, passengers and freight having to cross by means of a steam ferry.

The 'old' tunnel, subject of our brief study, was 6,025ft long (portal to portal) or 2.5 miles, including the 1 in 50 graded approaches. Of this, 2,290ft were under the river bed. Steam operation was slow and dangerous; trains had to wait for the gases of their predecessors to clear. The death toll of asphyxiated crews was grim: six men perishing in October 1904, for example. On economic and humanitarian grounds, something had to be done.

The solution was to electrify four route miles of line; the tunnel, its approaches, yards and exchange points totalled twelve track miles. The contract for electrification was signed in 1905 and the first electric locomotive passed through the tunnel three years later. Whilst they were about it, the Grand Trunk Railway (GTR) spruced up the tunnel, giving it electric lighting throughout, and replacing the encrusted grime of steam days with a bright whitewash.

The cost to the St Clair Tunnel Co was $543,000 (US) although in practice the bill was paid by the Grand Trunk Railway, owner of the tunnel company. The GTR was British-owned, with its headquarters in London. The GTR's investment was paid back within five years owing to improved operating conditions. Maintenance in a single year fell from $21,173 (steam) to $11,131 (electric). The GTR also saved considerably on carriage cleaning costs.

The St Clair electrification took place in, and was part of, the GTR's last, golden age of growth and expansion. More ambitious still was its participation in a transcontinental line, the Grand Trunk Pacific. This was hugely costly to build and was becoming problematic even before the Great War dried up the capital and immigration on which it depended. Ten years after the St Clair Tunnel electrification the GTR foundered, later becoming part of the Canadian National system.

A brief word on the men behind the electrification. The President of the GTR was Sir Charles Rivers Wilson, former imperial mandarin, Minister of Finance in Egypt, Comptroller of the National Debt (in the days it actually shrunk with ease), director of the Suez Canal, etc.[1] He was coming to the end of his term of office; long-term strategy as well as day-to-day details were increasingly in the hands of Charles M Hays, a rough, tough American railroader who succeeded Rivers Wilson in 1909.

It was he who pushed through the electrification, and the Pacific extension. There are signs that he began to doubt the wisdom of the Pacific venture – we will never know for sure: he went down with RMS *Titanic* in April, 1912.

The electrical mastermind was Bion J Arnold, an American. Ironically known as 'Father of the Third Rail' because of his pioneering work with that form of low voltage dc traction, he advised the GTR to go for the new, relatively untried, single-phase ac for its tunnel work. When the GTR asked for contract submissions, the Edison-General Electric interest came forward with a third rail, low voltage dc proposal, but the Westinghouse ac idea won the day in 1905.[2]

Arnold's advice was audacious, given the early and relatively untried state of ac traction. It was also sound; the installation ran with few problems for half a century. The unusual 0-Co-0 locomotives were high on reliability, low of maintenance charges and problems although, admittedly, they ran for short distances at low speeds.

Some technical details: the St Clair system was energised at 3.3kV, 25 Hz, the low frequency reflecting in part the 'state of the art' issues of early ac that included commutation problems and flashovers at higher frequencies. The voltage was also rather low by contemporary traction standards (London, Brighton & South Coast Railway, Midland Railway – both 6.6kV; New Haven, 11kV) but this is explained by the tight clearances in the steel-walled tunnel, and the danger of arcing at higher voltages. The mileage involved was so short as not to raise questions about voltage drop, or current boosting.

Power came from a generating station on the Port Huron bank of the river that supplied it with coal. The plant represented the latest thinking about such installations: automatic feeding of boilers by hoppers; forced draft by electric fans; two Westinghouse-Parsons turbo alternators, etc. There were four Babcock & Wilcox 'quick firing' boilers and a separately fired superheater. 'Obnoxious gases' as they called carbon emissions in those far off days were scattered to the heavens from a single, massive chimney.

Outside the tunnel the copper catenary (twin conductors, suspended from steel messenger wires) was hung from steel towers, 250ft apart. Inside the tunnel it was held by steel brackets and insulators fixed into the roof. Other beneficiaries of the power

Above left: Sir Charles Rivers Wilson. Centre: Charles M Hays Right: Bion J Arnold

Right upper: Baldwin-Westinghouse ac locomotives, permanently coupled in pairs. Dating from 1907 this Z-2 class lasted until 1959. They were equipped with Westinghouse 'diamond' pantograph, and accoutrements of the old order: wooden doors, brass bells and chime whistles. With a starting tractive effort of 33,000lb, 750hp per half-unit and a gear ratio of 16:85, these Co+Co machines could attain 35mph, but usually trundled along, c10mph and had an upper limit (enforced by a speed recorder) of 30mph.

Right lower: Typical St Clair Tunnel train in electric days, emerging at Port Huron, Michigan portal

system included YMCA establishments in Sarnia and Port Huron (110V), as well as tunnel lighting (480 bulbs) and some 30 arc lights in the yards at the terminals. An interesting feature of the dc arc-lighting was that the ac was rectified by early mercury-arc rectifiers, very avant garde for 1908.

The St Clair Tunnel Co locomotives were for slow, sustained slogging work. All axles were gear-driven by Type 137 Westinghouse 250hp motors and, in the absence of need for high speed; there were neither leading nor trailing wheels, resulting in good adhesion. Their design followed closely that of an experimental Westinghouse locomotive of 1904 which had demonstrated the possibilities of ac traction (but at 6.6kV in its case).

A considerable amount of compressed air was necessary for operating the St Clair electrics. Inside the box-cabs were air-cooled transformers and 100V motors (tapped from the main transformers) for driving the motor-cooling blower fans. More motor compressors, for braking, were located in the cabs at either end of each 'half unit' locomotive. In common with the standard Westinghouse approach (also used on the Mersey Railway, whose recent electrification had been Westinghouse-financed) control was electro-pneumatic. Compressed air also operated the bells and chime whistles; electric heat kept the sand boxes dry. These locomotives had a plain, boxy appearance, none too prepossessing, but be not deceived – they were as technically advanced as any motive power of the day.

By February 1908 all six of the new 66-ton locomotives were ready, as was the entire system. On Thursday, 20 February, locomotives 1308 and 1309 took a 700-ton freight train from Sarnia to Port Huron. With 'Superintendent Jones' at the controls, and Westinghouse resident engineer H H Rusbridge 'in attendance' the inauguration was entirely successful. Ideas about hauling mixed steam and electric trains were soon dropped: it required 20 minutes to dry off the tunnel insulators after one of the exhaling 0-10-0Ts had passed through. The new modus operandi, for a while, was electric operation for 18 hours per day; steam six. After May, 1908, the service became all-electric and the 'stretch tanks' were confined to yard duties.

The new system soon settled down. Westinghouse handed it over to the St Clair Tunnel Co in November 1908, with celebrations and junketings at the Hotel Vendome, Sarnia. By most criteria the St Clair electrification was a great success, journey times were cut and tunnel capacity increased by about a third. Claims were made that the system was the most dense electrical operation in the world, with twenty-six freight and fifteen passenger trains every twenty-four hours. Although twin-unit working was the norm, single, triple and quadruple headings were all employed on occasion.

There were a few minor motive power developments later in the electric era. A new locomotive, identical to the originals, was acquired in 1927. Two ex-Chicago, South Shore and South Bend electrics were added also in 1927. Perhaps most surprising of all, a petrol-electric (North American, 'gas-electric') was bought from the Canadian National (CNR) in 1941 (ex-Canada's National Harbour Board, ex-Montreal Harbour Commissioners) to assist with catenary maintenance when the current was switched off. This had been originally been built by

ST CLAIR TUNNEL Co
1305
1306

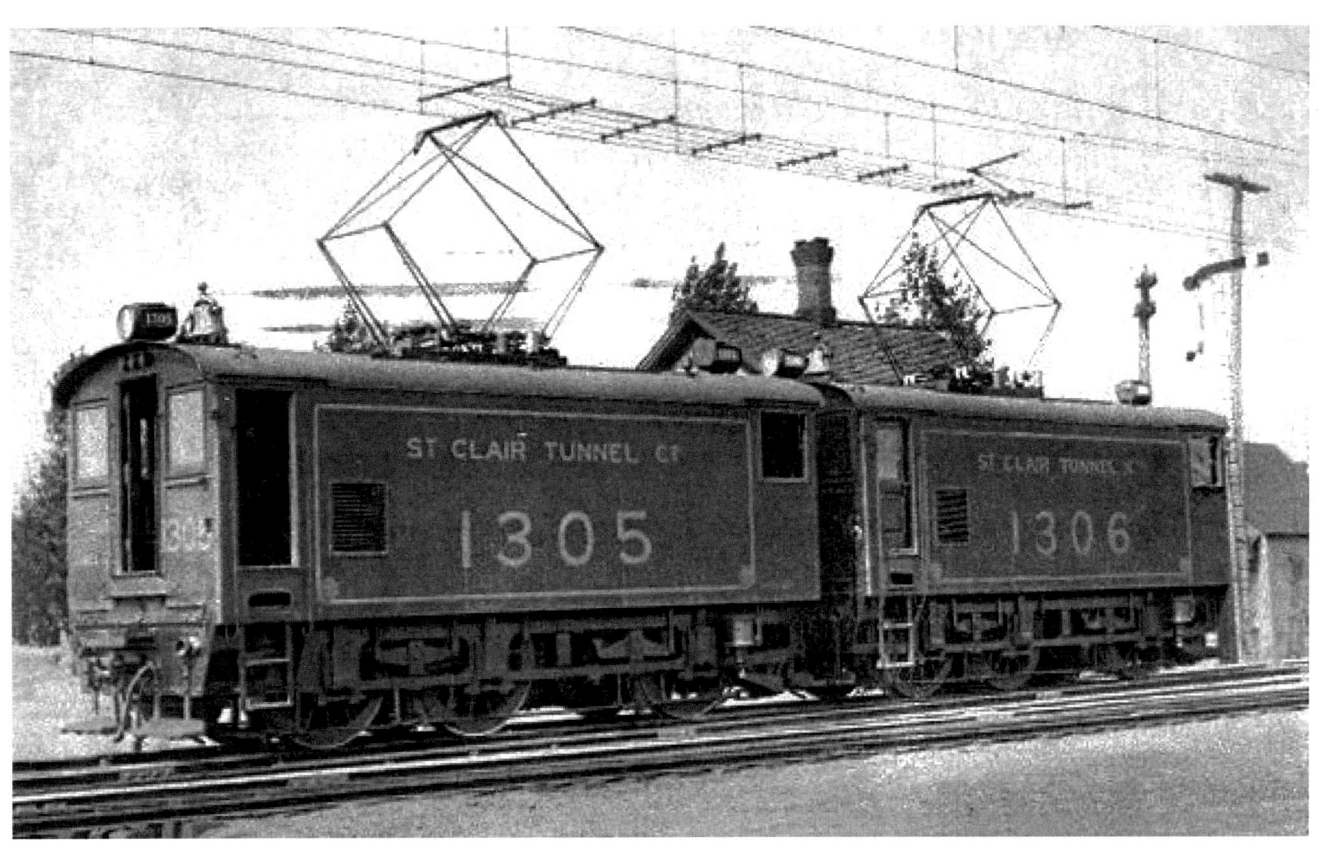

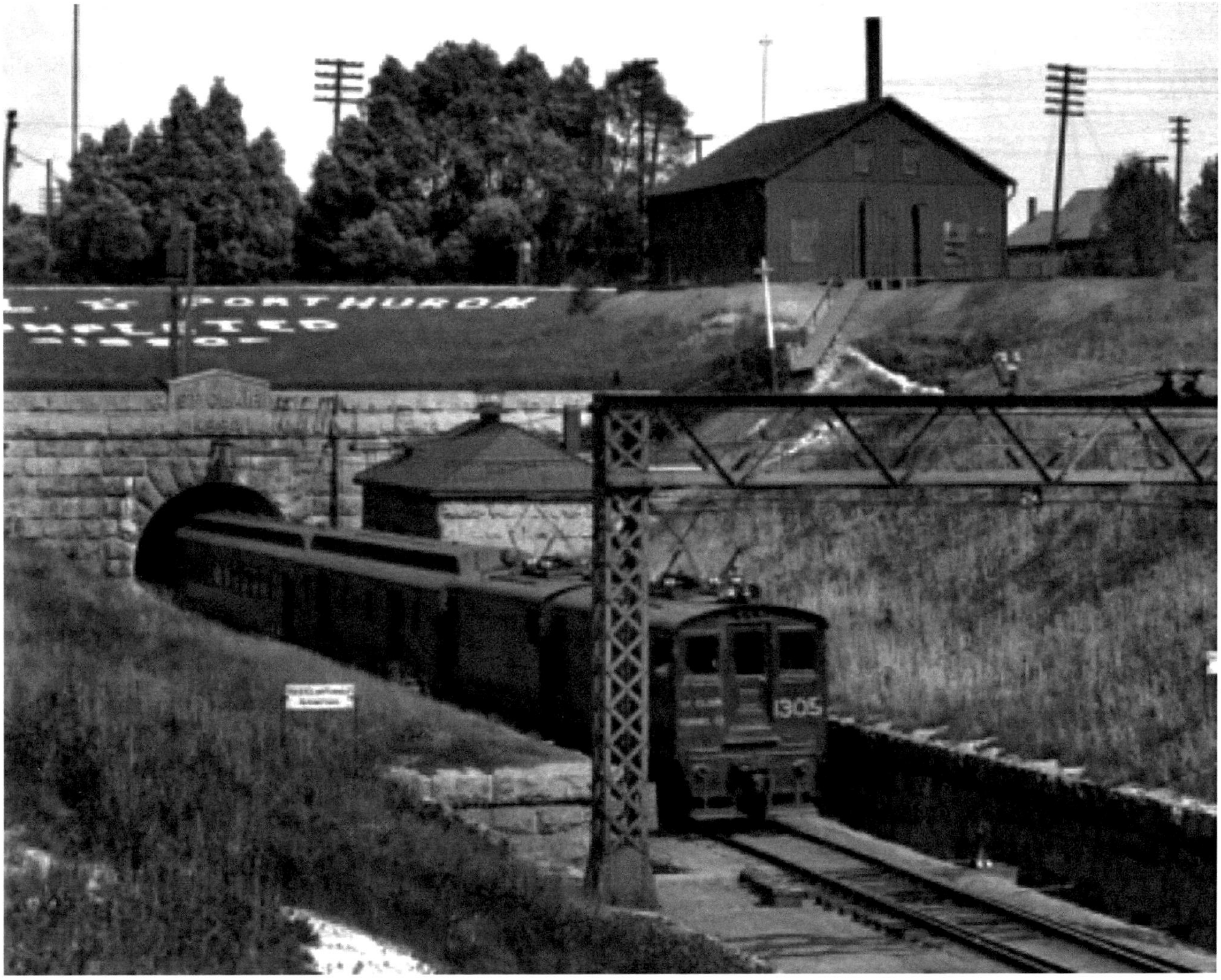
1305

Typical St Clair Tunnel train in electric days, emerging at Sarnia portal, Ontario.

English Electric and its official builder's photograph shows it in the familiar cutting near the Dick, Kerr works, Preston. After de-electrification, it was returned to the CNR. It was scrapped in 1968; its electric sisters had been withdrawn and scrapped in April, 1959.

This quiet piece of ground-breaking engineering demonstrated the way ahead. It was less well-known than the mainline electrification of the New York, New Haven & Hartford Rail Road, another ac pioneer, and was without many followers in the New World, although elsewhere ac was on the move; the St Clair Tunnel electrification was one of the first straws in the wind.

NOTES

1. Sir Charles Rivers Wilson (1831-1916) was very much an Establishment patrician (Eton & Balliol College, Oxford), a career civil servant and, later, financier. He was also chairman of British Electric Traction; as President of the GTR he held office 1895-1909.

2. Bion J Arnold (1861-1942): first made his name as mastermind of the 'Intramural Railway', a third rail system that served the World's Columbian exposition, 1893 (see Mike G Fell, 'Queen Empress, Chicago and Ben Robinson' in The Stephenson Locomotive Society, a Centenary Celebration, 1909-2009). He went on to supervise the electrification of the Interborough Rapid Transit lines, New York, and the New York Central's electrification of its lines leading to Grand Central, New York.

THE GENERAL ELECTRIC GEARLESS PROJECT

It reads like one of those 'railway trivia' questions: what ran over the Rockies, under New York, deeper still under London, but not through Northallerton? To save the reader wasting intellectual effort in unravelling this near-pointless conundrum, the answer is: GE 'gearless' electric locomotives.

The lines in turn were: Chicago, Milwaukee, St Paul & Pacific ('The Milwaukee Road'); New York Central; Central London, and the North Eastern Railway, of England.

But first, why 'gearless'? The very word implies that gears, of some kind, are conventional and to be expected, and yet this particular item is not furnished with them. The origin of the gearless concept lies, about a century ago, in the ancient rivalry between Westinghouse, the champion of ac electrification, and Edison, the dc stalwart. The Edison camp gradually coalesced around the General Electric Company, hereafter GE or 'General Electric' – no relation of the late-lamented British GEC.

In the early years of electrification, not only did the ac-dc war spice engineering lives, but there were, predictably, close searches for devices and processes that could be patented by way of excluding rivals from a burgeoning market.

The GE gearless motor grew out of this era of furious competition. The term can be readily explained. Rather than mounting electric traction motors either on the bogie or main frame of the locomotive, the general rule, although coming in many diverse forms, the gearless system simply mounted the armature of the motor actually on (i.e., around) the axles of the locomotive. The main frame carried only the field magnets.

This plain and straightforward answer to the problem of transmitting power to axles suffered a major flaw: it added greatly to the unsprung weight bearing on the axles, and could give the track, as well as the electrical machinery, a hard time. For all their complication, the geared systems were easier on the track since part, or all, of the weight of the motors was shouldered in sprung suspension of one kind or another.

There were attempts to mitigate the problem whilst keeping some advantages of 'gearless' transmission, largely by wrapping the armature around a tube, or 'quill' surrounding the axle which it turned by teeth-extensions bearing on driving wheel spokes. This ploy was tried by the Baltimore & Ohio, and New York, New Haven & Hartford lines, but this only revived complication.

The GE gearless motor was largely the brainchild of the company's Asa F Batchelder, who held four patents related to the design. Batchelder's tribe of gearless locomotives came on stream in the early twentieth century, but they were by no means the first gearless electric locomotives.

Around the time this chapter was first published in the Society *Journal* (July/August 2009) it had also included correspondence regarding the first main line electric locomotive of them all, produced by the Scottish chemist Robert Davidson, in 1842. From the fragmentary evidence available, it seems that his *Galvani* was probably geared, although its magnets, mounted rigidly on the floor of his device, partook of some of the attributes of the future 'gearless' type.

Most early essays in electric traction employed gearing of one kind or another, however. The gearless tradition, always a minority taste, reappeared in late Victorian times; certainly Siemens tried one out and later advocated the arrangement for the City & South London Railway, successfully as it turned out.

In the USA the Thomson-Houston Company (a constituent of the later GE) produced a two-axle gearless locomotive (strictly a 'Bo' machine according to the UIC notation) for display at the Columbian World's Fair, Chicago, 1893. It was later sold to a small industrial line, the Manufacturers' Railroad of New Haven, Massachusetts, where it soldiered on between 1897-1915. It has fortunately been preserved.

The heart of the matter: an armature-wrapped axle, New York Central c1906.

11

The PLM gearless experiment, with a battery *fourgon* **to the rear.**

In France, the PLM railway tested its first electric locomotive (E1, type 120) in 1897 by way of determining whether or not electric traction might be used for express trains. E1 was powered by batteries and had a 1-Bo configuration. The batteries were housed in a long bonnet-like affair behind which a slightly futuristic cab, with V-shaped windcutter prow, housed the controls. The locomotive was usually accompanied by its own *fourgon à accus*, a battery-housing van mounted on bogies.

E1 was described as having *un moteur à courant continu à transmission directe* (a direct current motor with direct, i.e. gearless, transmission). In top form the 360V dc apparatus generated 300hp and took four carriages (light ones, 50 tons the lot) at up to 60km/h on a down gradient, 40-50km/h on the rising one at Brunoy. Later, E1 shifted 102 tons at up to 62km/h. After further trials, it was withdrawn in 1905.

It was, however, in the USA that the gearless 'bipolar' motor found its home, for a brief season c1900-1920. Because of America's growing pre-eminence in electrical manufacturing, and also because of the powerful effects of American capital exports, it was to be the Central London Railway (CLR) that saw the first flourishing of 'bipolism', albeit a very brief and faltering one which left a mark that endures to this day on London's tube lines.

The design of the CLR gearless Bo-Bo locomotives has been accredited to both GE in the USA, and to Horace Field Parshall, the CLR's own consulting engineer. In later years (addressing the Institution of Electrical Engineers in 1916) Parshall gave credit for the idea of using locomotives, as such, to Sir Benjamin Baker. Quite possibly it was a shared effort. Its immediate inspiration may well have been the fleet of simple and uncomplicated 'Bo' locomotives operated by the City & South London Railway (CSLR) which had opened in 1890 with 13 gearless locomotives and one (No.2) with gears – but without much luck; it spent much of its life on light duties.

The 'gearless' design of the CSLR machines, however, differed from the typical GE approach by mounting the magnets at an angle to the axle-borne armatures, and mounting the whole ensemble, armatures and magnets, on to axles. Credit for this arrangement has been accorded to (variously) Edward Hopkinson, the line's consulting electrical engineer; Messrs Mather & Platt of Manchester, Hopkinson's employers; and Sir William Siemens, the unsuccessful bidder for the locomotive building contract. It certainly placed a weighty unsprung load on the axles (unlike the Batchelder-GE design), although the CSLR machines weighed only 10 tons apiece.

The CLR management had high hopes of its gearless locomotives, notably that being without gears; they would be a good deal quieter in the tubes than geared machines, then notorious for the whine of their machinery.

Alas, the CLR gearless locomotives were unsuccessful and short-lived, unlike the CSLR batch which put in over thirty years in some cases. There were good reasons for this calamity. First, the 28 GE-built CLR locomotives were much heavier than the CSLR ones (40 tons, to just over 10 tons) and secondly, they had a much lower centre of gravity, causing them to 'thump' the rails from the side as well as their top surfaces. But thirdly, and most serious of all, the massive gearless motors, or more strictly speaking, axle-mounted armatures, punished the track, and left serious corrugations. Heavy vibrations set up by this track-bashing could be felt in buildings lying over or near the line; complaints poured in and the Board of Trade ordered an enquiry.

The upshot was the withdrawal of the GE locomotives and their replacement by EMU stock, the UK's first use of this arrangement. The cars were built in the UK, but used American motors. The CLR had already been experimenting by putting gears on three locomotives – slight improvement – and constructing its first train with motorised coaches. The changeover occurred in the early summer of

1903, ending the UK's brief foray into the exclusive, if small, squadron of GE-inspired gearless traction machines – at least in practice, although as we shall see, possibilities were to return.

In spite of the indifferent record of the CLR locomotives, the true epoch of the gearless arrangement was only just about to start. It was presaged by a tragedy on 8 Jan 1902 in the smoke-filled tunnels leading to Grand Central station, New York. The driver of a train failed to see a red signal and smashed into a densely-populated commuter train. The accident did for 15 commuters, injured many others and caused public fury.

The NYC had already been contemplating electrification of its noisome approaches to Grand Central (rebuilt 1903-13), now it was hurried on its way by the New York State Legislature which decreed: no steam locomotives to proceed south of the Harlem River, w.e.f. 1 July 1908.

Advised by some of the leading engineers of the day[1] the NYC was quickly at work to meet the challenge. The upshot was some 60 miles of third-rail 600V dc electrification, two new power stations (Yonkers, Port Morris) and a fleet of unusual electric locomotives, the seemingly indestructible class S 'motors' built by American Locomotive Company (ALCO) and GE to Batchelder's designs.

These locomotives were 2-8-2, i.e. 1-Do-1 machines with four gearless motors apiece. Field magnets were mounted on the locomotive frame, armatures were wound around the powered axles: the classic 'bipolar gearless' design. These were the first electric locomotives fitted for multiple unit working. One of them by itself could shift 800 ton trains at 60mph; for short bursts they could generate 3,000hp.

It became the fashion in those far-off days to pit electric locomotives against their steam equivalents, by way of proving the obvious since a 'motor' was able to draw on the massive horsepower resources of giant power stations; the class K Pacifics (the chosen rivals) had to do it all themselves. The Pacifics did well, all things considered, but the motor (class S, 6000, unofficially *'Old Maude'* and still in existence) won. In spite of accelerating more quickly than the motor, the Pacific was soon overtaken and left well behind.

The NYC ordered 35 of the promising class T 2-8-2s, mainly in two batches. But, after so bright a start: disaster. A double-headed, electrically-hauled express came off the rails at speed only three days after the inauguration of the new service, killing 23 passengers – cruelly ironic in view of the ultimate cause of electrification in the first place. The cause of the accident remained mysterious, but to be safe the NYC rebuilt all of class T as 4-8-4s, or 2-Do-2s, after which they settled down to long, useful lives, re-equipped for lighter shunting duties in the 1920s.

Classes S and T became a byword for rugged reliability; the gearless motors gave little trouble; armatures could last over 30 years untouched in some cases. Prototype 6000 was preserved in 1965, at a hale 61 years of age.

The classification of the NYC bipolars is slightly complex. Pioneer 6000 became class T-1; then came the 34 motors of class T-2 (1906) and finally a dozen of class T-3 (1908-09). At various points before 1910, the 'T' became 'S', and there were 47 locomotives in all.

Contemplating the almost rustic, unsophisticated simplicity of these gearless motors one is minded of Panhard's description of the crash gearbox: *c'est brutal, mais ça marche* – it's rough, but it works!

In fact, the bipolars were more subtle than they looked. For one thing, the bugbear of unsprung weight was considerably reduced by mounting the field magnets on the main frame, or strictly 'motor truck' – unlike the CSLR variety that mounted complete motors on axles. The NYC motors had just less than 5tons unsprung weight on each axle.

Rustic simplicity: line drawing of the essentials of a New York Central bipolar, showing emblematic motors and stub pantographs.

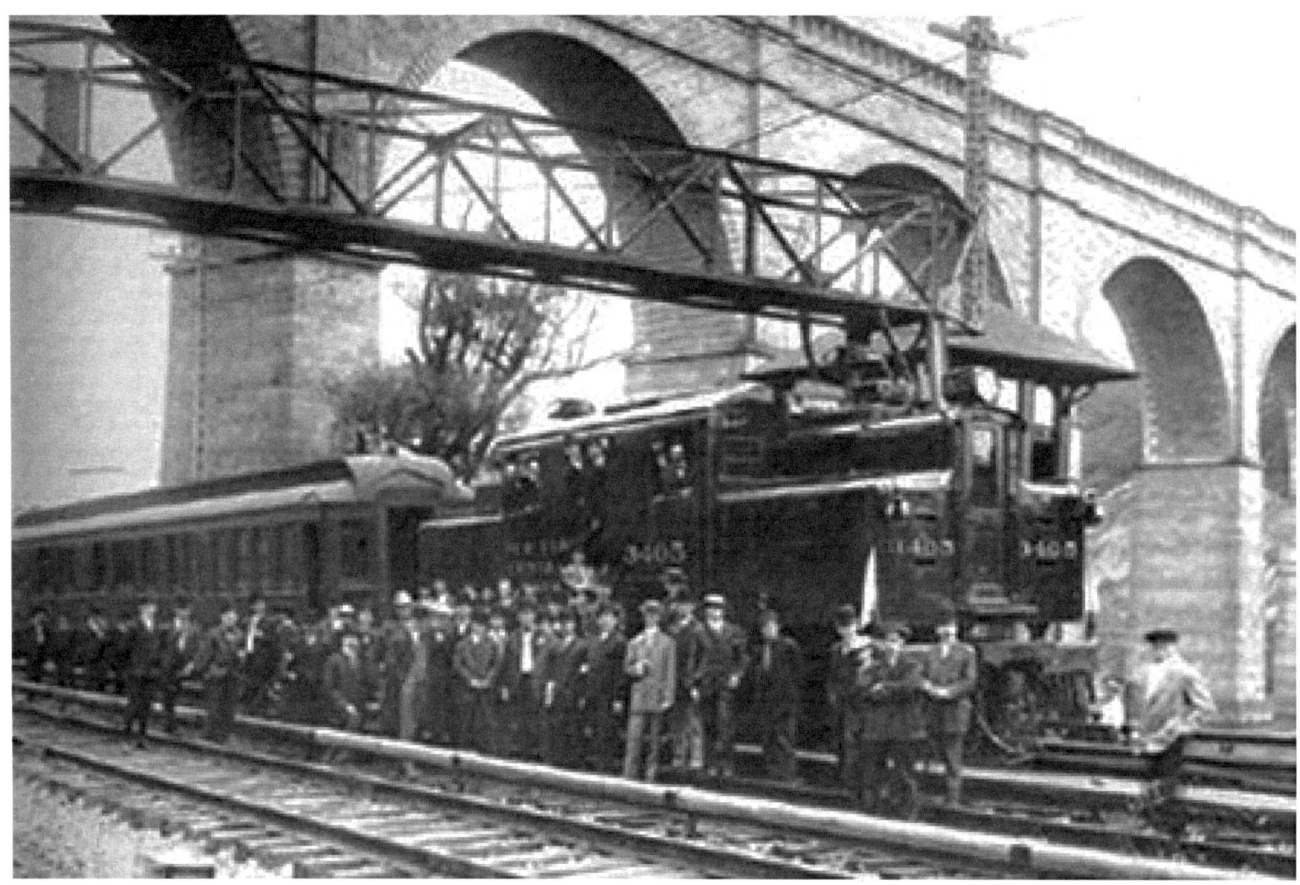

Demonstration train: Frank Sprague, inventor of the multiple-unit system stands beneath the 'T' of Central, front row, in this shot of the first New York Central train to be hauled by a bipolar gearless locomotive (3405) – 30 September 1906.

By employing only two field poles rather than the usual four on a dc motor, one on each side of the armature and of flat-faced, vertical configuration, the design was not only neat, but permitted slight but necessary travel of the axle up and down, to meet the irregularities of the track. The magnetic flux ran through all poles in series, returning via the steel motor frame. There was considerable elegance about these arrangements – and they ran, virtually problem free, for over seventy years in some cases.

To our latter-day eyes, the NYC bipolars may look a trifle quaint, but theirs was an honest 'form follows function' design. There was, for example, the curious front door leading to switchgear compartments, complete with handle, panel and glass pane. And then, atop the roof extension, the stubby, exiguous pantographs ('dc shoe' in NYC lingo) that picked up electricity at complex junctions where the third rail was hung aloft from massive gantries. The gearless motors themselves were unprotected from the elements, but they seem to have coped competently with all challenges, including both the right and wrong types of snow.

The cavernous cabs were entered by side doors, also panelled, and the driver had a fair prospect ahead, past a long clerestory roof extension bearing the headlight, warning bell and pantograph support. The world saw nothing like them before or after, but for well over half a century they almost defined Grand Central – and like all the gearless tribe, they were not only eco-friendly, but unusually quiet.

The last of the NYC bipolars, an S-2, was withdrawn as late as 1981. Three are preserved; not a bad record by most criteria.

The 'Batchelder era' of GE bipolar, gearless electric locomotives rose to its zenith shortly after the Great War, following some quiet years. Five huge bipolars were built for the Chicago, Milwaukee & St Paul Railroad[2]; designs were suggested for the UK's North Eastern Railway (NER), and a prototype was constructed for the Paris-Orleans Railway in France. After which, silence.

The 'Milwaukee Road' as it became known, was the last true transcontinental constructed in the USA (1909) – probably unwisely since its expensive extension to the Pacific coast barely ever paid its way and assisted in keeping the system insolvent for most of its existence.

The availability of low-cost hydro-electricity, the strong corporate interest of the Montana copper industry, and the successful high-voltage electrification of an important feeder line, the Butte, Anaconda & Pacific, persuaded the Milwaukee board to go for 3kV dc electrification. Two segments of the Pacific Extension were duly electrified: Harlowtown-Avery (440 miles) and Othello-Tacoma (on the Pacific coast) 207 miles, completed 1920.

In spite of their economic flaws, the two electrifications were a great technical and operating success. This was the world's first main line electrification of any great size, and it ushered in the era of

A Milwaukee bipolar in its classic form; copious sandboxes on the front beam.

high-voltage dc traction. As such it was much visited and studied, by Vincent Raven of the NER amongst others.

For fast passenger work on the Tacoma extension, the Milwaukee ordered five massive electric locomotives(1919), the 'bipolars' as they became known. Their wheel notation almost defies classification by the Whyte or UIC systems: 1+Bo+Do+Do+Bo+1; idler wheels at each extremity were 36in diameter, the twelve sets of driving wheels, 44in. In spite of their huge size, the EP-2 class, as they became, could hit 65mph; at 265 tons per locomotive.

The extraordinary plethora of driving wheels reflected the need to keep axle load down, especially given the problem of unsprung weight arising from the axle-mounted armatures. The locomotives were also 'triple articulated', i.e. in three distinct sections, to add to their flexibility – possibly mindful of early NYC problems with rigid electrics and a low centre of gravity. So, although the wheelbase was 67ft, its rigid portion was only 13ft 9in.

When going 'all out' an EP-2 had a starting tractive effort of 118,000lb, and a running tractive effort of 42,000lb, when it would be drawing 888amps. Regenerative braking could be applied in the 11-30mph speed range. Their happiest speeds were below 60mph; above that, in their later years, they could start to 'hunt' and suffer flashing in their gearless motors. Like other gearless locomotives, they were surprisingly quiet, although their air whistles were of the 'shriek' variety, rare in the USA, home of the melodious chime.

In appearance, the EP-2s were unique and unmistakable, with their central box section, a cab in three parts, shared out between each section, and massive semi-circular hood extensions fore and aft. It is a matter of taste, but one feels that in their first, unlined black guise they looked their unambiguous best. In later years they were tarted up in jazzy colour schemes, much admired and recalled by many, but on the pretentious side; 'mutton dressed up as lamb'.

Although the EP-2s were bi-directional, most driving was done from the 'A' cab since only it bore a kWh meter, an odd omission. This involved elaborate turning at either and of the line to keep cab 'A' in front. Like the NYC bipolars, the EP-2s had the slightly improbable front and back doors leading to such apparatus as switchgear, low-voltage motor-generator, etc.

Generally popular with crews, except possibly the 'fireman' who had to attend to the troublesome oil-fired train heating boilers, the EP-2s soldiered on reliably until the waning of passenger services. Refurbished none too effectively in the 1950s they had a short, less happy spell in the Mountain Section, suffering increasing problems of wear and age; they were then withdrawn 1958-60. One was preserved, a monument to American technology, also of a singularly eco-friendly kind, it so happens.

In spite of their relative success, neither the NYC nor the Milwaukee bipolars had any emulators, although they came close on two occasions. First, Britain's North Eastern Railway (NER). In the course of doing homework for the proposed York-Newcastle electrification, Vincent Raven (Chief Mechanical Engineer, NER) and Francis Lydall, of Merz & McLellan, the NER's consultants, made visits to the USA. Amongst the centres of expertise visited, GE Schenectady, New York was high on the list.

Having learned broad specifications from the NER team, GE sketched out a possible gearless design for the East Coast Main Line. It was of a 2-Co+Co-2 articulated design (see illustration). Merz & McLellan took the matter further, drawing

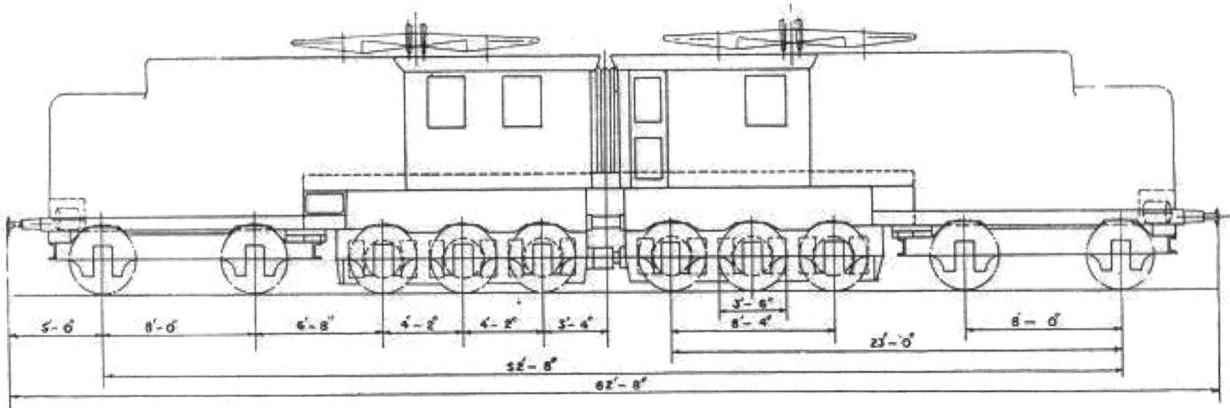

The GE outline sketch for a possible York-Newcastle bipolar gearless locomotive.

up plans for light and heavy versions, 2-Do-2 (4-8-4) and 2-Co-Co-2 ('double Baltic Electric'), respectively. At the same time, Westinghouse had also pressed their case, resulting in a 2-Co-2 geared design, the one eventually agreed by the Board and, of course, the No.13 of immortal fame (1920-50). The cautious directors could not agree to the production and trials of the GE gearless version, and so the world was not to see one shimmering through Northallerton, or negotiating the 'biggest diamond crossing in the world', the approaches to Newcastle Central. A pity, since one feels *Even the ranks of steam could scarce forebear to cheer.*[3]

GE had rather better luck with the Paris-Orléans Railway (P-O), however. The P-O had been amongst the world's railways tempted into the ac camp before the Great War. Afterwards, it had second thoughts, especially when the Milwaukee Road demonstrated the practicability of high voltage dc working. Also, according to its Chief Mechanical Engineer, M Bachellery (addressing the Institution of Electrical Engineers in 1924), the threat of inductive interference between ac conductors and telephone lines played a part in this change of mind.

Having decided to adopt 1.5kV dc, the P-O tried out some possible ways ahead. Hence its attraction to GE who supplied a singular bipolar gearless locomotive, constructed in Schenectady (1924), an articulated 'double Baltic, or 2-Co+Co-2, 601. After extensive trials, it was reconstructed by Schneider as a non-articulated 2-Co-Co-2 in 1929. Like the NYC machines, it had third rail pick-up shoes, and dwarf pantographs for the lines to the Gare d'Orsay.

The P-O locomotive was none too successful, its main fault being unsteadiness: *très mal conduire… malgré ses bogies directeurs*, according to Machefert-Tassin[4] – broadly, 'very bad tracking … in spite of guiding bogies'. The ghost of the NYC 2-8-2s seems to have returned, hence the later conversion to 4-12-4; the inability of the Franco-American bipolar to exceed 100 km/h (as opposed to the specified and hoped for 120km/h) without hazardous instability proved its undoing. Machefert-Tassin implies that the locomotive was essentially an electrician's offspring, not the ideal of a mechanical engineer.

Which rounds off the 'gearless bipolar' saga as far as the GE sub-species was concerned.

Taking stock The saga of the GE bipolars was an unusual one in the annals of locomotive history, shot through with paradoxes. A machine of rugged simplicity and ease of maintenance, on the one hand, exacted a price for this elsewhere, notably thumping the track and giving civil engineers problems. And yet, this problematic design produced locomotives that lasted for decades, even for generations, giving little trouble. The bipolars also seem to illustrate another of the odder manifestations of the 'geography of locomotion'. They worked well in their birthplace, the USA, but not elsewhere (compare Italian talent with Crosti boilers, German flair with diesel-hydraulics: yet sources of headaches when off their native turf).

Strangest of all, this unusual and exceptional design has been extraordinarily well represented in the preservation stakes. Virtually every type, except for the Central London and P-O sub-species; still exists, relative to their total numbers and other equally worthy relicts, even in abundance.

If preservationists were more pragmatic, what better than a slow trundle behind a bipolar gearless? Why recreate past steam forms when we have here an easily maintainable design that can be switched on five minutes before the first train, and then emits no greenhouse gases? Probably this is asking too much – and 650V dc third rails might give Health & Safety interests some serious jitters.

Meanwhile, there is the record of one of railway history's most and least successful designs. As with life, so with the bipolars: we have to live with contradictions.

NOTES

1. Including: Frank Sprague, inventor of the multiple-unit system of traction, and George Gibbs, mastermind of at least three major US electrifications and, it so happens, the constituents of the UK's Southern Railway. Scion of a well-off Chicago family, cousin of Alfred Gibbs, the great Pennsylvania Railroad steam Chief Mechanical Engineer.

2. Later '& Pacific' was added and, from the mid 1920s onwards its everyday brand name was 'The Milwaukee Road' – alas, gone with its electrics and most else.

3. Apologies, of course, to Macaulay, *Lays of Ancient Rome, Horatius, LX..*

4. See Y Machefert-Tassin, F Nouvion & J Woimant, *Histoire de la Traction Electrique*, Tome I, 1980, the standard general history. Also: William D Middleton, *When the Steam Railroads Electrified,* 1974, and F J G Haut, *The History of the Electric Locomotive*, 1969. Contemporaries interested in the bipolars: Philip Dawson, *Electric Traction*, 1909, and Edward P Burch, *Electric Traction for Railway Trains,* 1911.

THE OERLIKON ROD

The long history and taxonomy of pantographs and other arrangements for locomotives and rolling stock drawing electric current from conductors has received little recognition, at least in the UK: German and French treatment has been rather more systematic.

Current collector (or 'pickup') history falls into three overlapping periods. First, the pioneering era when a wide range of forms was tried out, with some evolutionary selection of the fittest. The second phase, the twilight of which still endures, was characterised by the dominance of the 'diamond' pantograph, in various forms for railways, and of trolley poles for a diminishing number of tramways. Finally, our own times, have seen the emerging dominance of the half pantograph, aka 'single arm' or 'Z' form typified principally by the Brecknell Willis, and Stone Faiveley types which can cope better with the high speed demands of the 21st century.

Confining our enquiry to overhead collectors, it seems that the simple trolley pole was the invention, mainly, of Frank Sprague in the USA. The diamond pantograph, quickly adopted by Westinghouse is often attributed to John Q Brown of the Key System (1903) operating near San Francisco. The other main category of early years, the bow collector, was largely a Siemens development from Germany. The UK saw most of these types, although on a much more modest scale than mainland Europe or the USA. The Midland Railway, for example, evaluated both the diamond and bow forms on its Lancaster-Morecambe-Heysham line, apparently finding that the diamond coped better with high winds off Morecambe Bay. The bow collector was also employed by a few industrial lines and the Sheerness tramways, where it received warm appreciation from the Board of Trade inspector. In later years various modified bows were used by (e.g.) the Leeds and Glasgow tramways. An even earlier type, the horizontal fixed bow, was designed by Edward Hopkinson and is used to this day on the Snaefell Mountain Railway, Isle of Man.

English usage in this field is a trifle imprecise. The bow collector, for example, is generally the 'lyre' on mainland Europe. The Germans, famously thorough in these matters, have a root for all types of pickup: *Stromabnehmer*, 'current extractor'. The particular rarity examined here, the Oerlikon rod or wand pickup was, therefore, a *Rutenstromabnehmer* – a 'rod current extractor'.

The Oerlikon rod or wand type did not make it to the UK, being confined to a few early single-phase ac lines, principally two in Switzerland, one in Sweden. It was the invention of the ac pioneer Dr Emil Huber-Stockar, chief engineer of Maschinenfabrik Oerlikon, (MFO) about 1904. It was highly ingenious in concept, a triumph of lateral thinking, but unfortunately not really adequate or sufficiently robust for the emerging world of heavy traffic and high speeds.

The rod collector was conceived as part of a package, the other part being the Oerlikon side-contact catenary. In practice the rod collector could oblige with current collection from conductors situated at the sides of a vehicle (the Oerlikon idea)

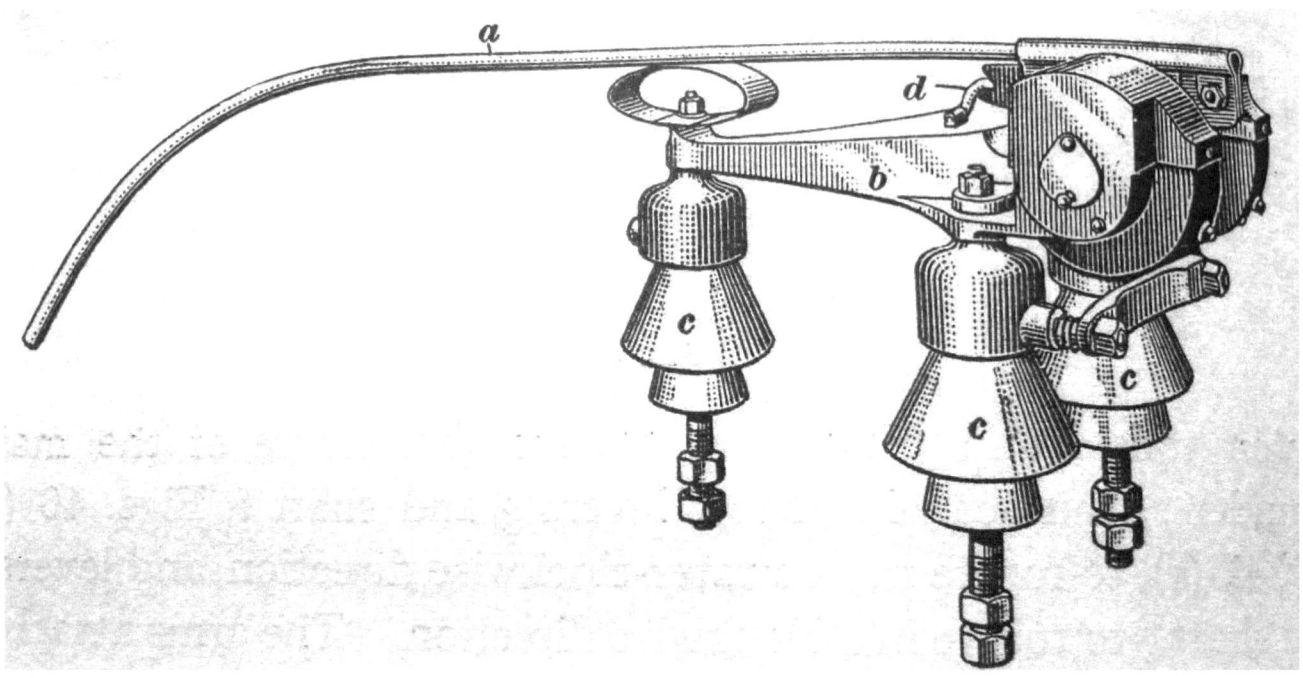

Diagram of the Oerlikon rod conductor from Dawson's *Electric Traction on Railways*, **1909.**

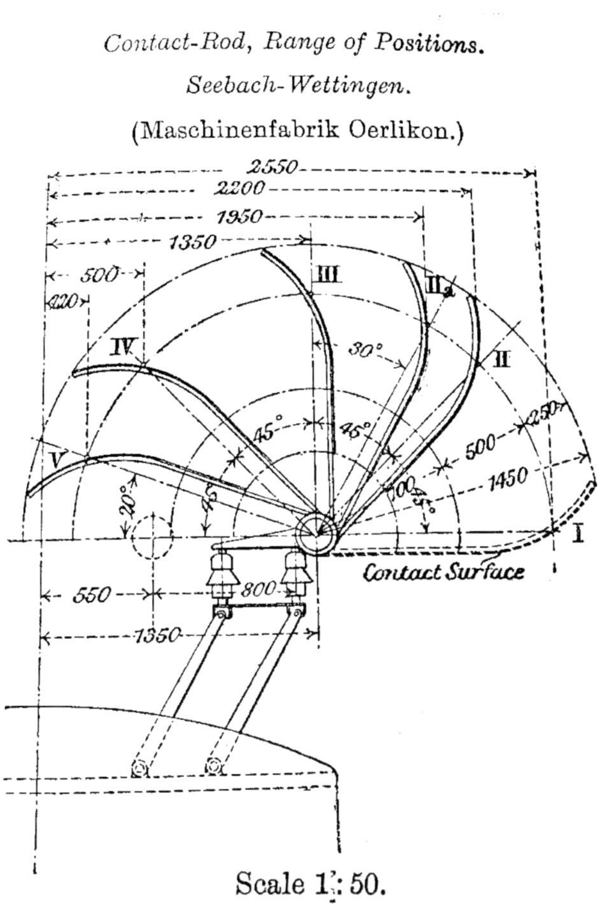

or above it, and at any point between, a perfect and continuous arc of contact locations. Oerlikon reasoned that a side mounted conductor wire could save material and complication vis-à-vis the orthodox overhead type, although in practice very few systems adopted it.

Where the writ of Oerlikon ran, the rod conductor appeared, but not elsewhere. It had three main locations: Seebach-Wettingen, Switzerland; Swedish State Railways; and the Locarno-Bignasco or 'Vallemaggia' line in Switzerland.

Seebach-Wettingen has an honoured place in the history of electric traction, being the first realistic, and sustained, attempt to use high voltage, single-phase ac as a source of motive power. The Oerlikon Company financed this experiment, effectively hiring a short line from the Swiss Federal Railways in the hope of persuading them to adopt this new arrangement. In the event, the SBB declined to adopt the system, although after a short interval and some politicking behind the scenes it relented, thereby establishing Switzerland as a major user of the Middle & North European standard of 16kV, 15 Hz ac, although Seebach-Wettingen ran on 15kV ac at first.

For its experiments, MFO first developed an experimental B-B locomotive with rod collectors (1904), duly tested on its private tracks. At first Oerlikon tried out industrial frequency, 50Hz, but with the state of electrical technology at the time it was found to be problematic, hence replaced by lower frequencies. Even so, it would appear that the industrial frequency 'Holy Grail', now the norm, was first supplied from Oerlikon side-contact catenary, collected by rods, thereby giving the Oerlikon rod collector an important symbolic place in the history of electric traction.

The experimental locomotive was soon transferred to the Seebach-Wettingen line. It was joined by two other experimental locomotives, another B-B, and later a Siemens Co-Co, both boxcabs. Rather like the contemporary three-phase ac in Italy, the Seebach line got extensive coverage in technical literature of the day, part of the search for the 'best buy' by engineers. Sir Philip Dawson, in his heavy classic *Electric Traction on Railways* (1909) noted of the rod collector: 'The collector, by its rotation, automatically adapts itself to any change in position of the conductor'. But, he added: 'Whilst for small units utilising but small currents, the Oerlikon form of collector bow would appear to have given satisfaction, where high speeds and heavy currents have to be dealt with... the wear... does not seem to have been as satisfactory as might be desired'. Dawson's conclusion was based on the findings of exhaustive tests held in Sweden which had compared the Oerlikon system with a Siemens 'lyre' and a more conventional diamond pantograph. The latter form was deemed superior, particularly at speed and in windy conditions; it was to be employed in Sweden

Upper Huber-Stockar's diagram of the Oerlikon conductor. (Not reproduced to original scale)

Lower: Swedish State Railways (SJ) electric locomotive No.1 on collector comparison trial near Stockholm, Oerlikon rods clearly evident. Stockholm, Oerlikon rods clearly evident.

19

Seebach-Wettingen locomotive No.1, rod conductors raised.

for much of the last century.

The weight of the rod and its actuating spring had a complimentary relationship, at least in theory. When the rod was in its initial position (see diagram) the coiled spring did all the work of pressing it up against the conductor. At the other extreme, the spring was at its weakest, but the weight of the rod on the conductor was at its maximum, although in practice one suspects that effectiveness of contact tended to diminish as the rod proceeded through its arc, not only because the force exerted by the spring lessened, but also because of the lightness of the rods, exacerbated by the small surface area of any actual contact they made.

Oerlikon seems to have realised the snag of variable pressure quite early. The *Elektrotechnische Zeitschrift* of 24 January 1907 showed that locomotive No.2 bore two modifications: duplicating the rods so that, in effect there was now one set for conductor wires on the left hand, another for the right. Also, both sets were now mounted on swinging arms that could extend the reach of the rods further to the left (or right) of the locomotive. But these improvements served only to complicate what had originally been a simple, relatively inexpensive system.

Huber-Stockar described his invention thus: 'contact was made by a rod swivelling about a horizontal axis parallel with and self-adjusting in a plane perpendicular to the direction of travel... '. He went on, describing his system to a meeting of the Institution of Mechanical Engineers held in Zurich, 1911, adding: 'A feature inherent in this system of contact line is the comparatively low level at which the contact-wire must be placed to fully realize the advantages'. But, the way catenary development was going rather vitiated his scheme for he added: 'The tendency now being to place the contact line high (6 to 7 metres) above the top of the rails, that system [the rod conductor and side contact] is not likely to be capable of more general application'.

And so it proved to be, although examples of rod-collection by the MFO system were to last into the 1960s, well within living memory.

Its three main examples were the three locomotives on the Seebach-Wettingen line; Nos 1, 2 and 3. Locomotive No.3 was an A1A-A1A machine from Siemens; Nos.1 and 2 (*Eve* and *Marianne*, respectively) were rod-drive B-B locomotives; both still exist in honoured preservation. All three locomotives had rod conductors, but also pantographs by way of comparative evaluation. About 9km of the line were equipped for the Oerlikon system, with another 12km of normal overhead catenary furnished by Siemens-Schuckert, hence the ability to evaluate two kinds of conductor suspension and two kinds of current collector.

The second example of the MFO rod collector system was a simple 0-Bo-0 locomotive for SJ, the Swedish State Railways where it was the pioneering No.1, built and run as a research locomotive with a different collector, much as on the Seebach-Wettin-

Locomotive No.2 of the Seebach-Wettingen experimental single-phase ac line in Switzerland, probably at Regensdorf-Watt station, c1907.

gen line. Swedish engineers seem to have concluded fairly quickly that rod collection had its drawbacks, particularly on lines with normal, vertically overhead catenary and conductors.

The Vallemaggia line, the third example of the MFO rod collector proved to be the longest lived example. It opened in 1907 using single-phase ac, 5kV, 20 Hz, a rather unusual arrangement, but of course this was a pioneer at the cutting edge of the unknown. The frequency was later changed to 25 Hz. Oerlikon side conductors ran beside the track at height of about 4.5 metres. Because this system was inappropriate for urban running it was relocated to a position directly the above the track in such locations with the voltage reduced to 800V, fed to the secondary windings of transformers on the cars. In spite of Huber-Stockar's hope that his rod collectors could cope with a wide range of positions, conventional diamond pantographs were employed in these urban surroundings so that the cars had to mount four current collectors; two of each type. Speeds were modest, a feature which may have worked in favour of this unusual example of the Oerlikon system.

The Vallemaggia route started with three motor cars operating with trailers. In 1911 MFO supplied the line with a rod drive 0-B-0 locomotive, equipped with rod current collectors and a large double-bow collector mounted on a long trolley pole for urban working. In later years the system was converted to 1.2kV dc, but still employing the Oerlikon rod conductor system, possibly its only venture into dc working. Part of a larger system serving Locarno, and linking with Domodossola in Italy, the Vallemaggia line closed in 1965, hence the comparatively recent survival of the rare MFO rod collector.

There may have been other applications and it would be interesting to learn their details from readers. Even so, it is a fairly safe assertion that the Oerlikon rod collector was a rare and unusual essay in the early years of electric traction, ingenious but found wanting, even in the eyes of its versatile inventor.

* * *

This chapter is an updated version of an article first published in the Journal of the Electric Railway Society, reproduced here by their kind permission.

KINLOCHLEVEN

A PIONEER HYDRO-ELECTRIC LINE AND ITS SETTING

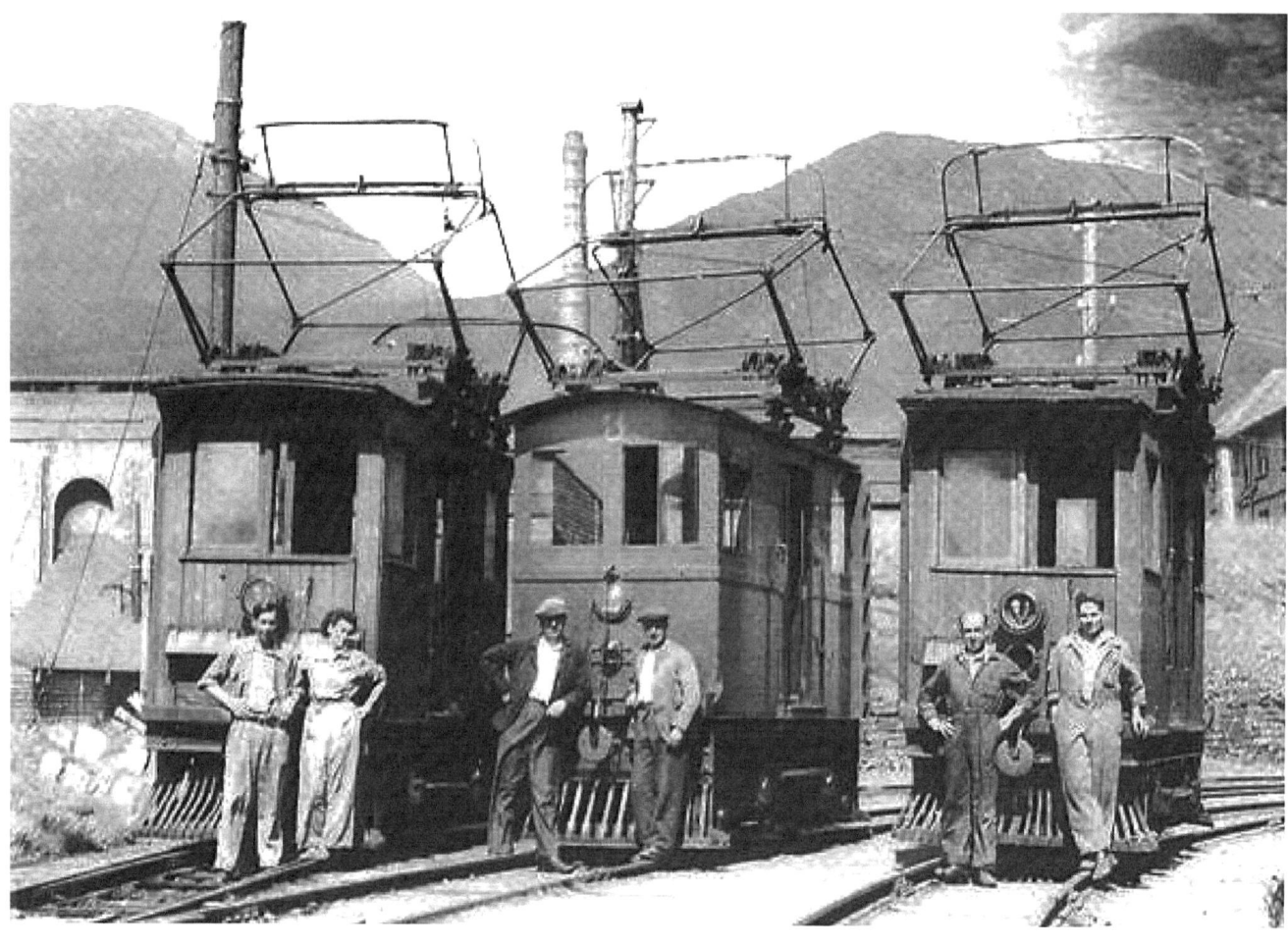

Kinlochleven motive power, c 1948, lined up near the smelters: Metrovick No.3 flanked by the Dick, Kerr twins of 1909 with attendant staff, including a stylish woman in working clothes.

A dreich day in the early 1950s, theoretically summer; but the windows of a MacBrayne's tour bus were damp from the showers outside as we dipped down to Kinlochleven, Argyll, slightly improbable but highly important metallurgical centre set in the Lochaber district of the Highlands. A brief glance of a narrow gauge railway, adorned with posts supporting conductor wires. Later, a small. teetering locomotive was spied in the distance hauling some wagons. That was all - your author's fleeting acquaintance with one of the UK's best-kept railway secrets, at least in those far-off days when railway literature was but a trickle of the flood we enjoy today.

Some visitors were more fortunate, able to inspect the installation closer up, even to ride on its singular locomotives and report the experience. And yet, even this small tramway-type line neatly exemplifies the wide scope of railway history: the many interconnections and ironies in its labyrinths. It touches, if occasionally briefly, on social and engineering networks, metallurgy, meteorology, technology and much besides - not to mention the perpetuation and correction of time-hallowed errors and 'urban myths', inseparable from the evolutionary processes of history.

Although Edwardian Britain was starting to slip, relatively speaking, in the development of new, electrically-based industry and commerce, there were some notable exceptions that demonstrated what might have been. One of these was the audacious British Aluminium Company (BAC) founded in 1894. It had already commenced hydro-electric smelting of aluminium in the Highlands with its Foyers scheme,1896.

Aluminium is a common element in the world's surface, but it occurs mainly as an oxide, its most workable form of ore being bauxite. During the nineteenth century, scientists had discovered how to convert bauxite into alumina and thence, by electrolysis, into aluminium - a strong, light metal resistant to corrosion, an excellent conductor of electricity, and therefore of great commercial potential. There was a major snag, however: the electrolysis was prodigal in its requirements for electricity.[1] In the 1890s there was no way in which the nascent British electricity supply industry could produce the abundance of

Building the railway near the jetty, c 1907: the telpher cableway to the upper works soars overhead, above a diminutive Dick, Kerr locomotive and various flat and open wagons.

cheap power required; it was poorly organised by a statutorily enforced muddle of local undertakings, contrasting unfavourably with its rivals.

The one possible source of cheap, electric power was hydro-electricity, hence the attractions of the Highlands to the BAC, already a company with strong Scottish influence - its consultant, later a Director, was Lord Kelvin, the sagacious physicist of Glasgow University; its chief for many years was W Murray Morrison, himself a Highlander.[2]

Following the success of Foyers, the BAC lit upon Loch Leven for its next and most ambitious project. Sea-going vessels could reach the upper end of the loch, although not quite as far as the proposed site of the smelters, hence the need for a jetty, wharves, and the electric railway to connect the various parts. A hydro-electric station could tap the abundant waters of Rannoch Moor by means of a dam across the Blackwater River, a fairly reliable source of water, given the rainfall in those parts, about 80 inches per annum at Kinlochleven alone.

Here the BAC built its aluminium works and a company town. A strong and, as they say these days, 'vibrant' community became established there, although the setting was not propitious - dominated as it was by melancholy grandeur, sombre cloud-capped mountains, frequently sunless skies, and much precipitation. It had an unhappy past: the nearby Glencoe massacre, government suppression of the clans, the Clearances, and depopulation. For these very reasons there were high hopes of the new plant at Kinlochleven[3] which, by 1914, was producing 90% of the UK's aluminium, much to be expanded by the needs of the Great War.

Early environmentalists reacted unfavourably to what they regarded as an industrial intrusion into the splendour of the Highlands, especially when the smelters got to work. Whilst it is true that a haze of smelter smoke hung over Kinlochleven, boosted by the reek of domestic chimneys, in a world of roaring tail races, whining turbo-generators, and an electric railway, no less, there were some gains. Kinlochleven did something to stem the tide of Highland depopulation, capitalism ironically healing some of the damage inflicted by the Clearances. Although no garden suburb, here was some modern

housing, clean and dry with running water, and, most famously, electrically-lit: some claim the first all-electric settlement in the UK[4]

The construction of the cyclopean Blackwater Dam and reservoir, four miles away and a thousand feet up in the hills, was on the heroic scale. Up to 3,000 navvies toiled in appalling conditions to build what was, for a while, Europe's longest dam. The main contractor, Sir John Jackson Ltd., fresh from building eight miles of the Manchester Ship Canal, successfully completed a 948 metre (half-mile) dam, 26 metres high (85ft), holding back a 13km (eight-mile) reservoir, eventually to contain 24,000 million gallons.[5] Although most of the work was manual, there were steam shovels, and an electrically-driven cableway brought raw materials up from Loch Leven. A contractor's railway linked the Loch with the upper works, involving two vertiginous, rope-worked inclines. Part of the Kinlochleven electric railway was, in effect, the remnant of a once longer, if short lived, system. The construction of these mighty works lasted from 1905 to 1909, over ran the intended time limit, and cost over the originally estimated £500,000 (about £50 million today).

The wharf and accompanying jetty for sea-going vessels and their attendant civil engineering work was carried out by a wholly-owned BAC subsidiary, the Loch Leven Pier Company. Its chosen subcontractor was Sir William McAlpine & Sons, thus the electric railway was constructed by the same contractors who had, a decade before, completed the West Highland Railway. The path of the railway had to be carved out of the narrow shelf where the mountains roll down to Loch Leven. This involved some long cuttings, and plenty of rock work near the jetty. Like the dam, much of the construction work was by manual labour, although McAlpines used a steam lorry and the narrow gauge line itself as work progressed. At the seaward end, the railway bifurcated; one arm ran on to the jetty that reached out into Loch Leven, the other served the waterside quay and wharves. Although successfully completed, the electric railway was part of a system that involved expensive transhipping, a process which led to its eventual demise. And yet it was necessary at the time, since the nearest railhead, Ballachulish on the Caledonian Railway (opened in 1903), lay at the other end of Loch Leven, cut off by the mountains, which made it a kind of fjord. The vestigial remains of the Stirling-Fort William military road that ran nearby could not have begun to support the swelling mineral traffic.

As the Kinlochleven scheme was launched in the twilight of the railway construction era, many people associated with it had strong railway connections, a theme worth pursuing briefly, since it demonstrates how all-pervasive railway engineering was at the heavy end of public works in the pre-1914 world. For example, Lord Kelvin, an enthusiastic backer of the scheme, had already invested in and advised the world's first hydro-electric railway, the Giant's Causeway Tramway in Ireland (1883).

The main hydro scheme, including the Blackwater Dam and reservoir, water pipes, penstocks, etc., was designed by Charles and Patrick Meik, who also

Left: Looking down the pipes which brought water to the turbines down below; a contractor's locomotive is just visible, far left.

Opposite: Down by the quayside: Metrovick No.3 and two wagons, Dick, Kerr locomotive beyond.

oversaw construction. The brothers, although born on Wearside, were Scottish by descent, sons of the eminent Edinburgh engineer Thomas Meik. Charles Meik had been apprenticed to Thomas Bouch, of Tay Bridge notoriety; perhaps because of this association he had then turned to engineering consultancy in Japan, whence he returned to join his brother in the construction of Port Talbot docks and their associated railway system in South Wales. In order to assist them on the Kinlochleven scheme, the Meiks hired another Wearsider, William Halcrow (1883-1958), later their partner, and whose name became famous in engineering and contracting circles, his firm being associated with the new Woodhead Tunnel on the Manchester-Sheffield line, the Victoria Line tube in London, and much besides.[6]

The main contractors for the dam and reservoir, Sir John Jackson Ltd., went on to build the Arica-La Paz Railway linking Bolivia to the Pacific Ocean in Chile. Jackson, another 'establishment engineer' of the time, specialised in massive harbour and waterworks, like the docks at Methil, Burntisland, Swansea and Barry as well as Dover Harbour, all railway linked.

The contractors for the electrical side of the Kinlochleven scheme, Dick, Kerr of Preston and Kilmarnock, were represented on site by Wilfrid Box who, as Resident Engineer, had charge of setting up the electric railway and the generators in the turbine house. Box (1878-1943), another Englishman (he hailed from Twickenham), already had important electric traction experience. He had been Resident Engineer for such Dick, Kerr contracts as the tramway generating stations at Portsmouth,

Grimsby, Singapore, Calcutta, and for the London County Council before he came to Kinlochleven. He left in 1911 to work on the Liverpool Overhead Railway for which he eventually became Chief Engineer and General Manager.

The water turbines which drove the Dick, Kerr dynamos came from the Swiss specialists, Escher, Wyss of Zurich. Their site engineer was Silston Cory Wright, a Yorkshireman, later to become a New Zealander. It was in his adopted country that his firm helped in large measure with the electrification of the Lyttleton-Christchurch railway (1928) and later still with electric rolling stock and locomotives for lines radiating from Wellington (1938-40). At Kinlochleven, Cory Wright had to see to the installation of eleven water turbines, two of which were dedicated to supplying energy for the town, railway and waterfront installations.

Even the architect engaged by BAC to design some of the industrial buildings such as the power house, as well as the senior staff houses, had railway connections: Alexander Alban Scott, who had at one time been employed by the Caledonian Railway and designed, inter alia, its station at Coatbridge.

Kinlochleven proved to be a prestigious 'battle honour' for ambitious engineers. It may have been perceived in an altogether more baleful light by the 3,000 or so navvies who worked in dreadful conditions on dam and reservoir construction, including relentlessly wet weather and an almost complete absence of a Health and Safety regime. Most of these men were Scots or Irish; site foremen had to be fluent in Gaelic as well as English. Their life was rough and tough - and occasionally short, a world of mudslides, casual use of explosives, cloying fatigue and, out of working hours, such distractions as alcohol-fuelled brawls. As a consequence it was hard to recruit and hold labour, one factor causing deadlines to be missed.

We know more than usual about this hard navvying life thanks to the literary gifts of Patrick MacGill, an Irishman from Co. Donegal, wielder of not only a shovel but also of no mean pen. Often referred to as 'The Navvy Poet', he wrote a classic account of the Kinlochleven works in his *Children of the Dead End, The Autobiography of a Navvy*.[7] MacGill's revealing statement that 'Only when we had completed the job... did we learn that we had been employed on the construction of the biggest aluminium factory in the kingdom' speaks volumes about the treatment of 'human resources' in the Edwardian golden age. The railway connection was manifest also in the case of MacGill who went on to platelaying and repair work on the Caledonian Railway between Greenock and Wemyss Bay before literary fame put his life onto an altogether different track, including positions as Librarian and Secretary to the Canon of St. George's Chapel, Windsor, active service in the London Irish Rifles, and some marginal work in the Hollywood film industry.

The sole purpose of the Kinlochleven electric railway was to convey raw materials inwards from the jetty and wharves to the main works, and to take refined aluminium ingots back to the Loch. For many years the line also brought in supplies. There was no passenger traffic or rolling stock, although BAC employees were allowed to travel on the locomotives.

The electrical side of the Kinlochleven railway fell into two conventional parts: motive power on the one hand, generation and distribution on the other. Dick, Kerr was the contractor for all of this. Power for the railway line came from two of the Escher Wyss water turbines driving two Dick, Kerr generators. The Pelton wheel water turbines were, in effect, vertically-mounted water wheels, stoutly built to cope with immense stresses brought about by a jet of high pressure water striking buckets mounted on the circumference of a wheel. They drove the generators at 400rpm and were rated at 75kW each. In later years, as traffic grew, a Brush generator was added, rated at 82.5kW and working at 800rpm. The generators supplied 500V dc straight to the overhead conductor wire suspended from long brackets, of shallow arc form that, together with the steel posts which supported them, were suitably covered in protective aluminium paint. The copper conductor wire, mounted directly in clips like those found on urban tramways, was of 0.75in diameter; where necessary the conductors were suspended from span wires.

In addition to powering the locomotives, the trolley wire also energised three travelling cranes, each of 3tons capacity, at the Loch Leven jetty. At first these had pole-arm collectors, also akin to those used on urban trams; one came from Stothert & Pitt, Bath in 1908, two more from Carrick & Ritchie, Edinburgh, 1916.

The railway itself was 1.5 miles of double, 3ft gauge track that, with sidings and a modest yard, totalled 4.22 track miles[8]. The line was laid on flat-bottomed rails, 65lb/yd; centre-to-centre of the double track ran to 9ft. The ruling gradient was 1 in 100, and the sharpest curve had a 50ft radius.

The mainstay of the Kinlochleven electric railway for its half-century or so of working was a brace of Dick, Kerr locomotives, joined later in the day by a third from Metrovick. Of these, more later; there were others that came and went. For example, when the aluminium works were expanded during the Great War, the contractor (Balfour Beatty) constructed a temporary 2ft gauge line served by two Kerr, Stuart 0-4-0STs, one of which eventually found its way to a whinstone quarry near Haltwhistle, Northumberland. Balfour Beatty also employed an 0-4-2ST on the 3ft gauge line. This was taken over by the BAC, later transferred to its works at Lochaber. Other birds of passage included a small diesel locomotive in 1939, transferred to BAC, Larne NI, and two battery electrics from Greenwood & Batley.

It was, however, the three diminutive -Bo- electrics with their outsize, beetling pantographs that almost defined the line. Two came from BAC's original electrical contractor, Dick, Kerr (Preston works) and were coeval with the system on which they worked, both arriving in 1908, both scrapped when the system closed in 1960. The third, outwardly virtually identical, although with steel rather than wooden bodywork, came from Metropolitan-Vickers in 1947, construction sub-contracted to E E Baguley Ltd.; it was also scrapped in 1960. The origin of this No. 3 has given rise to a species of 'urban myth' to the effect that all three came from Metrovick; not so, since the late, great MV of Trafford Park was only formed in 1919.[9]

At least one of the original twins was originally equipped with a cross-arm or scissors pantograph, a relatively uncommon version of the diamond configuration, in vogue from time to time - some of the early ac electrics of British Railways sported this form. Later, all the Kinlochleven electric fleet had orthodox diamond pantographs fitted, albeit with shallow horizontal bows fitted on top, resulting in an unusually large pantograph:bodywork ratio, brought about partly because the conductor wire lay 16-19ft above rail level.

Each locomotive was rated as capable of hauling five 6 ton loaded wagons at 10mph up the 1 in 100 gradient. Like some continental European electrics, they doubled, as noted, as passenger vehicles, offering sparse accommodation for up to six workers moving from the jetty to the smelter, or vice-versa. In their latter years, at least, the locomotives were painted green, a suitable hue for such an eco-friendly railway, running as it did on entirely sustainable energy.

The dimensions of the Kinlochleven four-wheelers were (rounded up): 14ft 7in length, 5ft width, and a bodywork height of just over 10ft. The wheels were 2ft 7½in diameter, each axle driven by a 35hp motor through reduction gears, 5:1 ratio, hence the 'unmistakable whine of an electric motor' noted by Dennis Gill back in the 1950s (see References below). Controllers also came from Dick, Kerr, their DK1 type with series-parallel control, and five notches for the electric brake[10].

Operations of this short line were straightforward; a typical incoming alumina would have consisted of five hopper wagons, more conventional wagons being used for supplies and outgoing ingots.

An electric railway of any kind was rare in Edwardian Britain, although electric tramways were becoming quite common in urban areas. It has sometimes been asserted that it was the first electric railway in Scotland, but this is not so - although it may have been better known than its rivals. It was beaten to the 'pole position' by at least two others, the Carstairs House tramway (1888-1895) and the Winchburgh Railway of the Oakbank Oil Co., the shale oil region of West Lothian, 1902. There were also two 'textile lines', that of the Levenbank Works (United Turkey Red Co.) near Balloch, c1904, and another serving the Almondbank Bleach Works near Perth, opened in 1911. So, although it is incorrect to claim that the Kinlochleven line was Scotland's first electric railway, it is not wildly wrong either; it was certainly one of a promising group of pioneers.

The Kinlochleven electric railway had a good innings, 1909-1960, its early locomotives and flat-bottomed track seeing it right to the end. From the 1920s, road traffic began to erode the monopoly of the electric line. A through road to the nearest railhead (Ballachulish, by then on the London Midland & Scottish Railway) had been started by the military, and some German prisoners-of-war during World War 1, enabling many normal supplies to come into Kinlochleven by lorry, without the necessity of transhipping to vessels. By 1960, BAC was converting virtually all transport operations to road transport, thereby saving further transhipping, and rendering the wharves and railway line redundant.

BAC itself had been at the receiving end of a particularly bitter hostile takeover in 1959. Its new owners set about 'increasing efficiency' - a phrase known to many job-losing victims of the phrase then and later - in its Scottish operations. The railway was accordingly shut in 1960 and removed. Interestingly, the new chief of BAC, Sir Ivan Stedeford, was at that very time grappling with the growing problem of what to do about the staggering financial losses of British Railways, at the request of the Government. He was also grappling with his fellow-investigator, Dr. Richard Beeching, whose arguments won the day, and whose outcome we know well - a further increasing of efficiency. Even if he did not get his way with BR, it was on Stedeford's watch that the Kinlochleven line ended its days.

Kinlochleven, 1933 - town in foreground; jetty, wharves and railway terminal beyond. The new Ballachulish road that eventually did for the railway snakes up the mountainside to the left. Photo: *Valentines No.JV-223480,* **courtesy University of St. Andrews, Department of Special Collections.**

Another of history's ironies is that at one time BAC owned another electric railway, albeit not at all in the Kinlochleven mould: the Chemin de Fer Martigny-Osières in Switzerland, the remnant of an unsuccessful attempt to set up hydro-powered aluminium smelting in those parts, not far from the St. Bernard Pass. Although this popular tourist railway still operates, BAC itself is with our yesterdays, as is its Kinlochleven initiative. The smelters had become relatively small and costly by the time of their closure in 2000. Kinlochleven itself endures, however, skilfully reinventing itself as a centre for tourism and other enterprises. The turbine house remains, part of its output feeding the Grid. Well worth a visit to see the trackway of the unusual electric railway where, for a while, mineral trains plied their way beneath the mountains; there is much else to experience including the windswept moors and lochs which saw it all come and go.

NOTES

1. It required, in those early days, 30,000kWh to produce one ton of aluminium. Four tons of bauxite converted to two tons of alumina.

2. Later Sir William Murray Morrison (1873-1948), who had oversight not only of Kinlochleven but also similar schemes at Foyers and Lochaber, as well as Stangfjord and Kristiansund in Norway, a leading metallurgist and engineer of his generation.

3. An Anglicisation of the Gaelic Ceann Loch Liobhann. Fortunately the BAC forebore to name its new town Aluminiumville, seriously suggested at one time. The Gaelic sounds a good deal better.

4. Electric firsts, like railway firsts, are unstable and contestable: we are at least fairly certain of some: first house lit by hydro-electricity, Cragside, Northumberland; first town with electric street lighting: Godalming; first electrically-lit church, Blockley, Worcs., now Gloucestershire.

5. In those days it would have been 24 'milliards', a billion meant a million-million, but like much else in culture, this word came to ape American usage.

6. Including advising Barnes Wallis on the 'Dambusters' initiative and Mulberry Harbours.

7. See especially the chapter 'De Profundis', a useful corrective to those who complain loudly about H&S regimes. MacGill (1889-1963), whose *Children of the Dead End* first appeared in 1914, wrote other accounts of hard, working life in Scotland, tatty-howking (Anglice,'potato picking'), drain digging, tramping and platelaying, etc.

8. But Gill, op cit, stated the total route length to be 1,666 yards - in 1957.

9. The otherwise meticulous and scholarly Industrial Railway Society 'Handbook N', *Industrial Locomotives of Scotland*, lists all three machines as 'MV', a source later quoted elsewhere.

10. No reference describes these brakes clearly: rheostatic? regenerative? magnetic? It would be interesting to learn.

11. This chapter first appeared as an article in *BackTrack* in May 2014 and is reproduced here by kind permission of its Editor.

REFERENCES

C E Box. 'Scotland's First Electric Railway', **The Electric Railway**. Vol.5. No.51, May-June 1950.

C E Box. *Liverpool Overhead Railway*, 1959.

Dennis Gill, 'An Electric Railway at Kinlochleven', *Railway Magazine,* Jan.1957.

Guthrie Hutton. , 2012.

'Our Wullie'. 'Loch Leven Electrics', **Railway Bylines**, Vol.8 No.2, Jan. 2003.

Andrew Perchard. **Aluminiumville ,Government and Global Business in the Scottish Highlands.** 2012.

Hamish Stevenson. 'British Aluminium in the Western Highlands', **Industrial Railway Record** No.206. September 2011.

THREE-PHASE AC TRACTION IN THE AMERICAS

Advertisement from Ganz, Budapest, a major player in the three-phase initiatives at the time. 1909, from *Electric Traction on Railways* by Philip Dawson, a man more strongly associated with single-phase ac, as on the London, Brighton & South Coast Railway's electrification which he masterminded.

Just over a century ago, various 'Battles of the Systems' were exercising engineers. The best known, the struggle between alternating current (ac) and direct current (dc) for domestic and industrial supply was won, in the main, by ac with which we now live.

A related field of contention caused much debate amongst the railway and tramway electrifiers. Broadly speaking, they had (c1900) three choices: first, dc which had proved straightforward and robust, easy to control, for mainly short-run systems, tramways and suburban railways. But, it employed low voltage, usually about 600V and, once any appreciable distance was involved, it needed expensive substations to boost the current.

Secondly, just on the horizon, but proving itself feasible, was single-phase ac, being pioneered in Switzerland and Germany. It was to have a great future for mainline work, particularly because it could be transmitted at high voltages requiring few substations. And yet, it too had snags including the necessity, at the then state of the art, to employ low frequencies (usually 25Hz, or 'cycles per second') of a kind that were unrealistic for domestic supply, partly because of the resulting flickering, stroboscopic effect on lights - hence the need to build and run special low-frequency systems for ac railways. Also, the ac motors of the day were on the weighty and sluggish side when compared to dc.

Third, came the false dawn (as it turned out) of three-phase ac (Italian, *trifase*, German, *Drehstrom*) full of early promise but, with the most notable exception of Italy, little adopted for railway work. It, too, had its pros and cons. The rugged simplicity and relative lightness of three-phase ac motors were offset by the problem of the fixed speed characteristics of the three-phase motor.

Most early systems could only offer two speeds, x and 2x, although eventually up to four could be arranged by pole changing. This situation had its attractions for timetablers who knew that a given train would soldier on, up hill and down dale at (say) 25mph. Operators saw the issue differently since bursts of acceleration to make up time, for example, were out of the question. Additionally, most famously of all, the system required three conductors; usually two overhead wires and one running rail, leading to fiendish complication at junctions, and complex, rather weird, pantograph and boom arrangements on locomotives.

Because Italy was quick off the mark with railway electrification it paid the price often exacted on pioneers, that is of adopting a system that was quickly overtaken by superior ones: hence the well-known Italian three-phase ac electrification, the standard model for all such installations.

Starting with the experimental and successful Valtellina line (1902) the Italians proceeded to

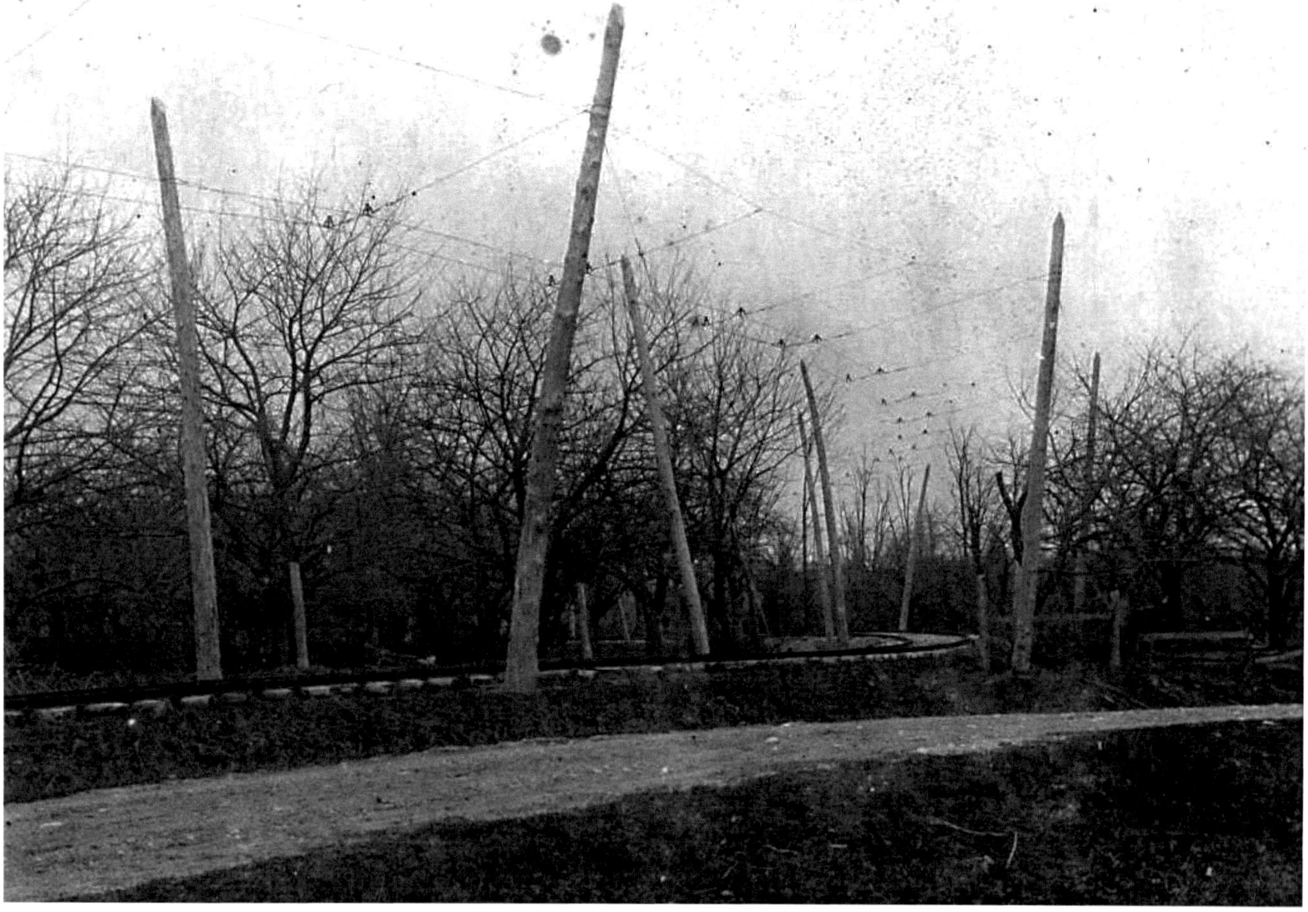

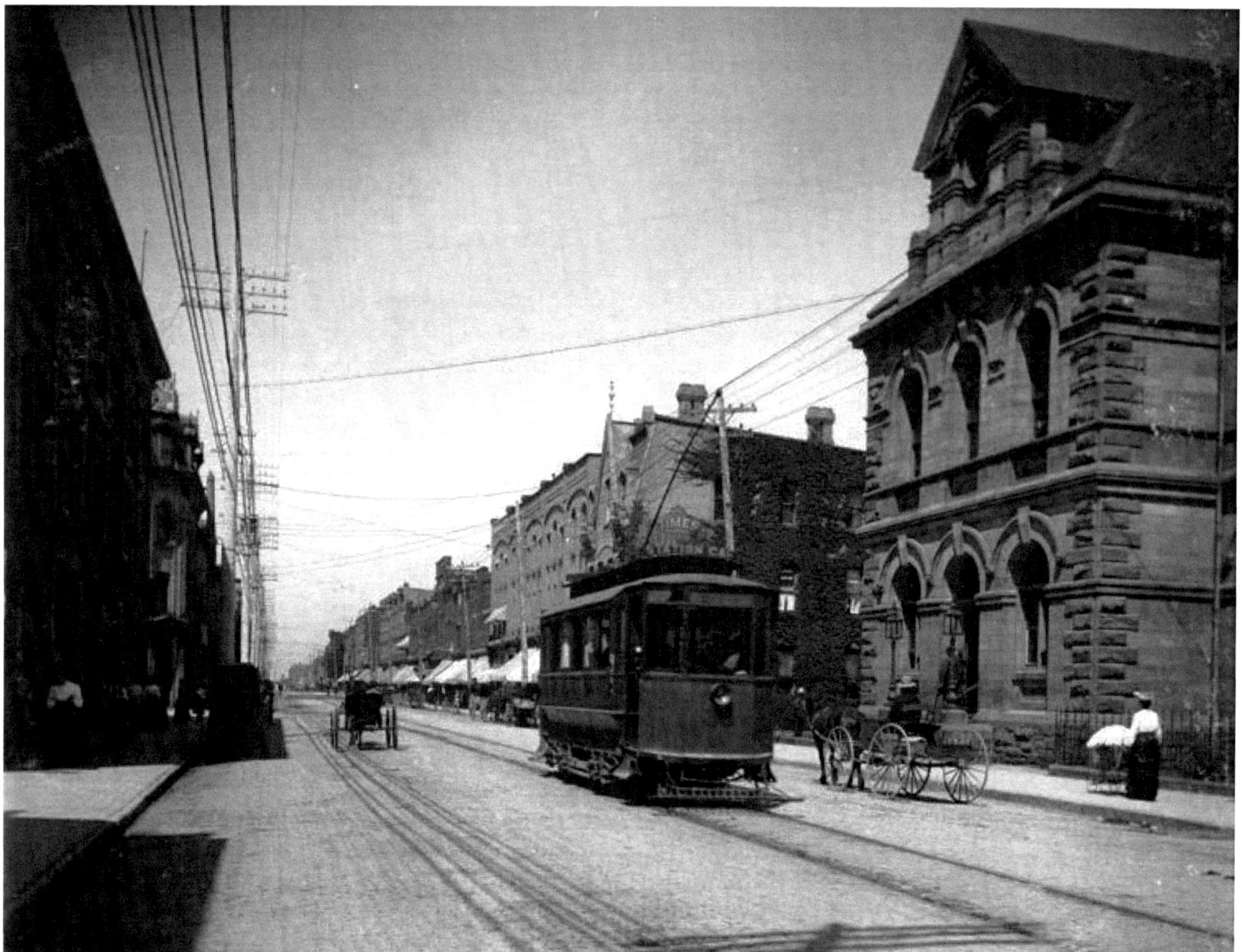

electrify, eventually, some 250 miles of main line with their *trifase* system. This was mainly confined to the North, notably the Ligurian coast, but there were outliers, up the Brenner Pass, and from Rome to Sulmona, a short-lived trans-Apennine line energised at 45Hz. The excellent and reliable Italian system, gradually replaced by high voltage 3kV dc, lasted well within living memory, until 1976.

Precisely because it was so extensive and closely studied, especially in its early years, the Italian three-phase system has dominated the literature; its coverage includes some notable works albeit mainly in Italian. Because of this Italian dominance in the three-phase field the many other, admittedly much smaller, three-phase systems have tended to be shunted into the margins of railway history. This chapter is a small contribution to their reinstatement as a significant part of the saga of railway electrification, offering some definite 'lessons to be learned'. Although much of their story lies in the past, in fact some systems continue to operate most successfully. They include, for example, two of the world's best known tourist lines, climbing the Jungfrau in Europe, and the Corcovado in Brazil. The story has an unexpectedly happy conclusion in that three-phase ac traction is gradually becoming the world norm, but as an outcome of very different arrangements to the old twin-conductor, fixed speed systems. The secret has been to develop reliable and flexible frequency-changing systems on board locomotives, and thus the ability to enjoy the advantages of three-phase ac without its snags.

Moving from North to South of the Americas, one found almost the full range of three-phase types: an Interurban railway; a main line; a mountain climbing rack railway, and a unique one-off, the Panama Canal ship-shifting electric 'Mules' for which reason the Americas are a good starting point for any world survey of this unusual system.

In Canada, the Southwestern Traction Co. (later the London & Lake Erie Railway), one of the least known electric railways, let alone of three-phase ac lines, was an unsuccessful foray of the Edinburgh electrical engineers, Bruce Peebles, into electric traction.

This firm had obtained a concession, from Messrs Ganz of Budapest - at the time the world's leading advocates and experts in the matter of three-phase ac traction - for exploiting their system in the UK and throughout the then British Empire. They were to have two such attempts to press the system on railway operators, the Southwestern Traction Co. and, in the UK, the Portmadoc, Beddgelert & South Snowdon Railway, later to be part of the Welsh Highland line. Neither initiative paid off; both were to cost Bruce Peebles dear, leading to serious financial difficulties.

The Southwestern Traction Co. was backed by another UK enterprise, Canadian Electric Traction. It changed its name to the L&LE in 1909. Bruce Peebles, which had a financial interest in the line, was duly awarded the contract for building and equipping it (1904). Advised by the Edinburgh firm, the SWT was consequently to operate on the 'Ganz System', i.e. three-phase ac. It was opened in stages, London-Talbotville in 1907 and the next year, through to Lake Erie, 28 route miles.

Records of the system are sparse. It presented numerous teething problems with overheating motors, serious arcing, and, worst of all, the need to run its line through St Thomas, which already had a low voltage dc line along the chosen route. Ganz claimed to have developed a system whereby three-phase ac motive power could run for stretches on dc. It was not successful, and in any case involved laborious raising and lowering trolley booms at either end of the town so as to run for a short distance on direct current.

The directors soon got cold feet over the ac system and decided to convert it as soon as possible to dc. They had only just placed orders for replacement motors when a calamitous fire, probably caused by a short circuit, broke out in the 'car barn' destroying much of the rolling stock. The line was eventually reopened as an orthodox dc route. Still none too successful, it closed as early as 1918, its 'three-phase phase' having lasted only about one year.

The British cars were built by Brush of Loughborough with Bruce Peebles equipment. They had twin trolley poles at either end, being otherwise unremarkable in appearance Interurban (or 'Radial' in Canadian terms) cars with monitor-type clerestories. Three were combination cars geared for 22mph and three were passenger saloons. In addition to their electrical problems, their bodywork tended to flex at speed; in short, not one of three-phase's most successful achievements.

Left upper: Photos of first generation, three-phase, Southwestern Traction operations are very rare; this rather poor quality item shows the twin conductor wires, typical of the Ganz three-phase system, strung up above the interurban, or 'radial' in Canadian terminology; the SWT was always referred to locally as 'The Traction'.

Left lower: The Achilles' heel of the SWT was its requirement to run under low-tension dc conductors through St Edwards, Ontario. The equipment of the day was not up to coping with this complication for high-tension, three-phase ac arrangements; here is one of the local dc cars of the St Edwards Street Railway outside the Post Office in 1905, just before the SWT came.

Great Northern box-cab locomotives for the three-phase Cascade Tunnel electrification. An unusual feature was the use of trolley-poles to pick up power from the conductors which varied considerably in height.

Further south, in the USA, the Great Northern Railway ran an altogether more successful three-phase system, one of the very few main line uses of the arrangement outside Italy.

The adoption of three-phase ac by the GNR nearly started a fashion for similar work in North America, but, in the event, did not. The Denver & Rio Grande RR seriously considered copying the GNR's initiative, even to the design of proposed locomotives. The Southern Pacific, urged on at the time (1908) by its controlling magnate, Edward Harriman, also considered three-phase traction for its mountain climbing stretches over the Sierra Nevada. Neither plan, alas, bore fruit.

The GNR electrification was short but mightily effective; an attempt to ease working over the congested and expensive route in the Cascade Mountains, Washington. The Cascade Tunnel (2.63 miles or 4.23km, and on a 1.7% gradient, West-East) had been opened by the GNR in 1900, eliminating eight switchbacks in the Cascades, 600 miles west of the Rockies. Steam working had led to horrendous conditions; smoke infested tunnels, long delays for freight and passenger trains, and lethal fumes lurking in tunnels. The GNR accordingly installed the USA's only three-phase ac electrification, opened for operations on 10 July 1909.

This was an entirely self-contained system run on environmentally friendly hydro-electricity, supplied at 6.6kV, 25Hz from a plant thirty miles away, transmitted at 33kV. Four box-cab locomotives were constructed by ALCO and General Electric, Schenectady, to take care of traffic, although they usually hauled or pushed trains through the tunnel complete with heavy steam locomotives that generated just enough power to shift themselves. Soon after electrification, the GNR discovered that their new electrics could manage 2,000 ton trains with ease, a great advance on the smoky pyrotechnics of its articulated 'Mallets' that had managed traffic until then.

Because the overhead conductors in the tunnel were of great simplicity, without complex catenary suspension, the box cabs, uniquely, had long tramcar-type trolley poles, two at either end, well able to cope with the considerable range in height of the trolley wires. Current was reduced on the locomotives to 500V by air-blast transformers for nose-suspended motors; fixed speed was 15.7mph (25.12km/h). Later, GE converted the locomotives to 'cascade control' (the usage is pure coincidence) whereby they could also run at half speed by changing the setting of motor induction windings, allowing all four locomotives, at times, to move a train without draining the power plant.

On one occasion, the driver of a locomotive was unaware in the gloom of the tunnel that his machine was, in fact, attached to a stalled train. The synchronous three-phase motors pressed on relentlessly, giving the impression of movement, but actually

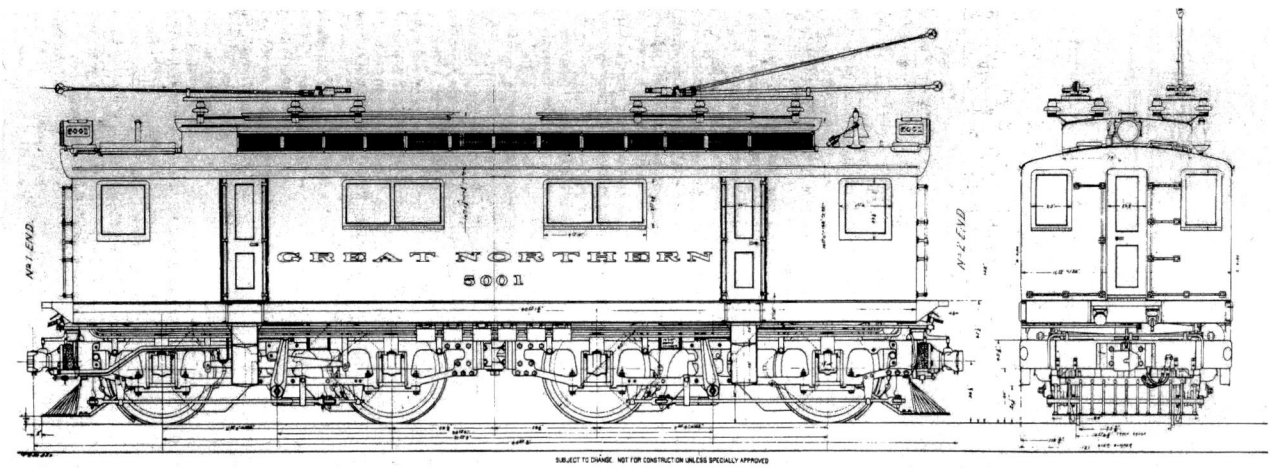

Above: Bo-Bo locomotive with body and running gear from ALCO, electrical equipment by General Electric, for the GNR (USA) - plain and unadorned, a representative of the shoebox school of bodywork design, a double-ended box cab typical of its era. Four GE-1506 induction motors producing 1,500hp for each locomotive, driving 5ft wheels. Rheostats, used in motor control, were located on the roof.

Below: A first generation Panama Canal 'Mule' hauling a US warship through the lock system; twin cabs, single windlass amidships, power picked up from energised rails in a conduit. These locomotives came from GE, Schenectady, weighed 43 tons each, and slogged on for over forty years - rugged simplicity and longevity being typical characteristics of three-phase motive power.

causing the web of the running rails to be ground down by about two-thirds.

This dramatic incident and a spectacularly fatal avalanche in 1910 apart, the GNR electrification was a notable success, allowing for the limitations of early three-phase systems. Because the system was so short, it was unable to employ regenerated current to assist climbing trains; this was generally dissipated by heating a water rheostat back at the power plant, somewhat denting the experiment's otherwise green credentials. Plans to extend it a further 91 km bore no fruit. It was eventually replaced in 1927 by a single-phase ac line, and a longer, lower tunnel. In its early years it received considerable publicity, but, lacking emulation, and being overtaken by single-phase ac and high voltage dc, it spent much of its life in historical shallows.

Further south came one of the most extraordinary rail-borne installations, let alone a three-phase example, anywhere on Earth - the Panama Canal lock-hauling railway 'Mules'.

Before looking at this unique system, a brief word is needed about American electrification. Like the Cascade Tunnel line of the GNR, the Panama initiative came in the golden age of US railway electrifications; the era when 18,000 miles of electric Interurbans were constructed, of the New York electrifications of the Pennsylvania, New York Central, and New Haven systems. This early and inspiring lead was lost after the Great War, with Europe taking over the baton. Why? Largely because big scale electrification (including many three-phase examples) often requires government backing in one form or another, not the American style at all [2].

Above: The second and third generation 'Mules' for the Panama Canal Commission were fairly similar; here is one climbing the stiff gradient to a Miraflores lock; Mitsubishi Heavy Industries design and build; 1961 onwards - working on 440V ac, three-phase, using the standard American frequency, 60Hz. Steeple cabs face inwards, with good fore and aft vision, but blanked off to the rear, convenient alfresco seats for yard crew, and two winches for more precise handling than the original single winch GEs. Photo: *Jean Paul Boule*

Left: The SS *Ancon*, first vessel to make an official (rather than actual) through voyage on the Panama Canal being assisted through locks by three-phase mules, photograph possibly 1914.

Ironically, the Panama system owed its very existence to public policy and finance, the entire Canal Zone and its working being the outcome of state policy, domestic and foreign. The original 'Mules', and the system they served, were firmly property of the US Government, as even their bodywork lettering made clear: 'US' inscribed at first on their sides. In recent years, however, the 'Mules' have been the responsibility of Panama, and its own Canal Authority. Formally known as towing locomotives, or 'Lock Mules', the simple 'Mule' description was accorded early and has stuck, even to the subsequent two generations that replaced the originals.

The main purpose of the 'Mules', once the canal opened (1913), is to ease ships through the series of locks (Gatun, Pedro Miguel, and Miraflores) principally by steadying them in midstream as they proceed; forward motion of vessels is aided by ship's engines. Hence the 'Mules' were equipped with massive winches powered, like the rack-driven rail system, by hydro-electricity generated by the Canal Authority, which could take advantage of the different Atlantic and Pacific sea levels to drive water turbines. The capstan drums could pay out or take in cable at a rate of 200ft per minute, protected by slip-drums that eased potential shocks arising from sudden jerks. Driving these 'Mules' requires great skill, given the tight clearance (as little as 2ft each side in some cases) of vessels which were subject to unpredictable and powerful eddying currents in the locks.

At Miraflores and Gatun there were a number of locks for the 'Mules' to negotiate, three for example at Gatun. Here the 5ft (1524mm) tracks swooped up (or down) on a 1 in 2.3 gradient, duly negotiated by the rack and pinion drive of the locomotives. Typically, four locomotives were assigned to a vessel, concentrating on steadying work, occasionally after a spot of haulage to help get a ship under way.

The first, and longest-lived, generation of 'Mules', forty in all, were constructed by General Electric, Schenectady NY, and delivered in 1912-14. There were slight differences, barely discernible, between the Atlantic and Pacific end 'Mules', although basic characteristics were the same: each 43-ton locomotive had two 220V three-phase motors, running on a 25Hz frequency current, delivered to two conductor rails in a conduit beneath the tracks, with a third phase in one running rail. There appear to have been some longer, articulated locomotive experiments, but the standard, well-known, design was as supplied by GE. Basic speed was 2mph, reducible to 1mph by 'cascade control'. Higher speed could be obtained by a mechanical gear shift. Outward appearance was curious, each 'Mule' having a cab at each end with the winch drum in between.

These famous locomotives worked on until they were replaced in 1964 by a new generation of 82 steeple cabs constructed in Japan by Mitsubishi. These were in turn augmented by a third generation of 'Mules', also from Mitsubishi, (1997-2002), all still working on three-phase ac and with a price tag of just over $2 million each.

The comparative data of the three generations of 'Mules', expressed in metric terms, demonstrates the march of electrical technology, in short 'more with less' in that the third generation of 'Mules' weighed less than the second, but had similar power. Up to eight 'Mules' may now be necessary to shift and control the largest vessels. The canal itself has been widened and improved, with new lock systems that may be able to dispense with 'Mules', although such is the volume of traffic that existing locks are still heavily used, thus ensuring the continuing reign of the 'Mule' dynasty for the time being.

Construction	Weight	Pulling Capacity (kN)	Speed (km/h)
1912	43 tons	111	3.2 - 8
1964	55 tons	311	4.8 - 14
1997 series	50 tons	311	4.8 - 16

Furthermost south of three-phase traction systems in the Americas is the Corcovado Railway in Brazil, another long-time survivor from the classic era of the three-phase, and, like the Panama Canal system, considerably modernised. It now caters for over 600,000 visitors each year. By a neat irony, the Corcovado brings us back to Canada, since its electrification in 1910 (750V ac, three-phase, 50Hz; 60Hz later) was carried out under Canadian auspices, altogether much more successfully than the London & Lake Erie enterprise and its predecessor.

Also, the Corcovado ran on electricity generated by hydro power like other lines under consideration here. More irony still; the Southwestern Traction Co. (London & Lake Erie), situated in Canada, one of the chief global locations of hydro power, was actually powered by a coal-fired steam generating plant.

The Corcovado (Hunchback, 2,300ft; 710 metres) is one of the famous eminences dominating Rio de Janeiro, the Sugar Loaf (Pao de Assuacar) and Tijuca being others. Nowadays, and since 1932, it has been crowned by an immense statue, Cristo Redentor, (Christ the Redeemer), the materials for which were brought to the summit, laboriously and over time, by the Corcovado Railway.

The 3.8km, metre gauge, Corcovado has been a rack railway (Riggenbach system) since its inception; first proposed in 1882 and opened by the Emperor of Brazil (Dom Pedro II) in person, 1884. It was rightly seen as a symbol of modernisation, backed by not only the Emperor, but also one of the country's most dynamic modernisers, F P Passos, later Prefect of Rio.

At first it was necessarily steam worked, by locomotives from Baldwin, USA, but in the early twentieth century the Canadian capitalists were at work spreading the 'gospel of electricity', first in São Paulo, later in Rio de Janeiro. Their chosen vehicle was a great utility company known locally as the 'The Light', strictly the Rio de Janeiro Tramway,

Above: A Corcovado locomotive and coach pausing at the Curva do 'Oh'! (see text)

Above Right: Paneiras station, Corcovado Railway, in the days of the Oerlikons.

Below left: Corcovado train in the 1920s, assembled tourists and Cariocas, smartly turned out at the summit, in front of an original Oerlikon three-phase ac locomotive of 1910, with twin trolley poles for picking up two of the three-phases.

Below Right: A Corcovado three-phase train with an Oerlikon locomotive of the first electric generation beneath the world-famous statue of Cristo Redentor, one of the principal magnets for Rio tourists. Photo: *Eduardo Coelho,* **in Rodriguez,** *A Formaçao.* **(see refs)**

Light & Power Co. Ltd. (Cia de Carris, Luz e Força de Rio de Janeiro, Ltda.). It set about purchasing Rio's mule-powered tramways and electrifying them with vigorous speed. The Corcovado, much run-down and barely operable, albeit on the up after a local takeover, was taken over in 1908 and electrified two years later. Power came from the 'The Light's' Ribiero des Lagos hydro-electric generating station, opened originally in 1906, supplying 6kV ac transformed to 800V ac for the railway at the line's own substation located at Painswick.

'The Light' chose the Swiss electrical engineering firm, Oerlikon Maschinenfabrik, to design and construct the three-phase locomotives. The Swiss connection was maintained some seventy years later when SLM-Brown Boveri built the next generation of motive power (two-car EMU sets) for a much modernised Corcovado Railway. Each original locomotive had two 155hp motors and weighed 15.5 tonnes - the passenger cars, weighing 4.5 tonnes, carried up to 66 passengers each. The second generation stock consists of seven cars, four motored, three trailers, weighing 19 and 17.8 tonnes respectively, and carrying variously 58 or 63 passengers.

The electrified Corcovado was an immediate success; running costs fell, capacity increased, and tourists could enjoy their climb to the summit free of smoke and smuts. In the past, typically, tourists and the local 'Cariocas' would arrive at the Corcovado for their 20-minute trip to the summit by the Aguas Ferreas bonde (tramcar) from Rio Branco Ave to Cosme Velho station, alighting there for the rack railway. The run to the summit had two stops, Sylvestre and Paneiras; en route the three-phase ensemble threads thick forest, but also sweeps out on to the edge of the Corcovado, presenting vast panoramas of southern Rio to the passengers. Amazed at the sight, their traditional exclamation has resulted in the naming of an unusual railway feature, the 'Curva do Oh!' - i.e. the Oh! Curve - although perhaps well-familiarised crews are more blasé about the phenomenon. The service is stepped up in the Summer months (January-March, southern hemisphere), but quieter at other times.

Fortunately, the Corcovado Railway, now in Brazilian ownership, flourishes still, continuing to operate on three-phase ac. Service is maintained by its four 2-car trains, neat vehicles in strong red livery.

The Americas may only have offered four of the world's admittedly few three-phase ac lines, but this brief survey may have demonstrated that they covered virtually the full range of operational possibilities: one main line, one Interurban, one rack railway, and an entirely unique one-off – the canal haulage 'Mules'.

They had some elements in common: all sprang to life in the period c1904-12, roughly the 'Edwardian era' when decisions about suitable, let alone optimum, forms of electric traction still raged on, unsettled. Three-phase ac was a player at the time, soon to be shouldered away by single-phase ac, and by dc. It was to return with a vengeance, but not quite in the way envisaged by its advocates of a century ago.

Readers will also have noted that, with the sad exception of the SWT in Canada, modest, constant speed was no drawback to the lines in question, and regenerated power had definite attractions, not the least being the saving of a fortune on perpetually renewed brake blocks. The long-life survival of half of the systems studied suggests that the engineers got it more or less right. They were also ahead of their time in another way: three of our four systems ran on sustainable hydro-power, also able to employ regenerative braking, sustainable options to the modern way of thinking.

A modern three-phase ac EMU set on the Corcovado rack railway, constructed by SLM/Brown Boveri, 58-seater motor car in front.

NOTES

1. Although the famous Pennsylvania RR New York-Washington-Philadelphia electrification did, in fact, receive Federal assistance in the 1930s, and public finance has propped up commuter lines in modern times, as well as the Amtrak system of long-distance passenger traffic.

2. This chapter first appeared as an article in *Locomotives International* No.92,Oct-Nov 2014 and is reproduced here by kind permission of its Editor.

REFERENCES

The standard work on three-phase traction is Giovanni Cornolò and Martin Gut, *Ferrovie Trifase nel Mondo, 1895-2000*, an exhaustive world survey, albeit with minor gaps.

Canada: John F Due, *The Intercity Electric Railways in Canada*, and also an excellent website: http://www.trainweb.org/ swtractionco/

Great Northern: William D Middleton, *When the Steam Railroads Electrified*. See also http://www.trainnet.org/Libraries/Lib001/CASCADE_GN.TXT

Panama: Cornolò and various contemporary commentaries, for example: Fred Talbot, *Electrical Wonders of the World*, 1925.

Corcovado: Cornolò, but also Allen Morrison, *The Tramways of Brazil*, and Duncan McDowall, *The Light, Brazilian Traction, Light and Power Company Limited, 1899-1945*, and H S Rodriguez, *A Formação das Estradas de Ferro no Rio de Janeiro - O resgate da sua Memoria*. The history of the line was covered in detail by Fred W Harman, 'Rails to Corcovado - (Estrada de Ferro Corcovado)' in *Locomotives International* No.79.

AN ODD VOLTAGE

THE RISE AND FALL OF 2.4kV DC

The French pioneer; Thury-designed Bo-Bo locomotive *Le Drac* for the C de F de La Mure, 1903: one bow collector contacted +1.2kV dc, the other −1.2kV dc. Photo: *Musée de Vieux, Geneva*

One of the youthful joys of closely observing trains was becoming familiar with locomotive classification. The 'ABC' for the ex-LNER fleet, with its logical and precise taxonomies, was a particular favourite. But, if one finds steam locomotive classification intriguing, spare a thought for the electrical specialist who has to cope with ac, dc, a huge range of voltages, frequencies, distribution systems and locomotive drives: nose suspended, rods, Scotch yokes, Buchli, quills, Bianchi – the list stretches away like a catenary over the plains.

Hidden away in the long evolution of electric traction are some rarities, for example unusual voltages. One such was 2,400 volts direct current, 2.4kV dc. Only three main lines employed it, all in North America. It was part of a quickly developing age of transition from the relatively cautious but popular and widespread 600V dc, via 1.2kV dc (something of a minority taste) to the eventual dc victor, 3kV that was adopted on a massive scale in the USSR, Italy, Belgium, etc. It marked a point at which manufacturers – exclusively General Electric in this case, were becoming more confident about higher voltage dc systems, as well as designing motors that could cope safely and reliably with increased tensions.

The chief advantage of raising the voltage was, of course, that it cut the number of substations required, and employed a smaller gauge copper conductor, with consequent savings on catenary support.

The 2.4kV saga was not, however, exclusively American. Some excellent Bo-Bo locomotives were supplied for it by English Electric and Beyer Peacock, in the relatively short era when the UK was a net exporter of heavy electric traction. Also, there were some minor specialist applications of this unusual voltage, one in Germany, one in Italy, and one in Chile, of which a little more later.

Looking back, nearly a century from its inception, one finds that the most significant aspect of the 2.4kV chronicle was, however, its ramifications – what it tells us about the wider scene within which it was located.

Before embarking on the main 2.4kV saga and in the interests of historical accuracy and fairness, the first (and now the oldest) line running on 2.4kV dc is a tourist and heritage line in France, the 30km, metre gauge St-Georges-de-Commiers – La Mure railway (C de F de La Mure, near Grenoble), dating from 1888 and electrified in 1903.

Its version of 2.4kV was highly unusual; it ran two conductor wires, one at +1.2kV dc and the other at −1.2kV dc. A short stretch had a single 2.4kV conductor, and the whole line was so standardised in 1950. This ingenious system was masterminded by a Swiss engineer, René Thury and was (and is) unlike any other 2.4kV line.

Its electric locomotives and railcars have come, variously from Thury; St Denis; Secheron; Pinguely; Ateliers du Nord; French Thomson-Houston (now Alstom); Chantiers de La Buire; and Brown Boveri. The first five locomotives, designed by Thury, were the world's first high voltage dc electrics. Because of the gradients on the line they had three braking systems: hand, vacuum, and rheostatic.

The main line 2.4 kV story starts in remotest Montana, on the poetically named Butte, Anaconda & Pacific RR. The BA&P started business in 1892. Its main purpose was to connect copper mines at Butte with a massive smelter at Anaconda some 26 miles away. In 1911 control of the system was acquired by the Anaconda Copper Co, a firm closely connected to the electricity industry. It decided to electrify the entire 70-mile line, using electric locomotives in place of its fleet of Consolidations (2-8-0) and Mastadons (4-8-0) to shift heavy ore trains, mostly composed of 50-ton double bogie hoppers.

The key figure in a complex web of ownership and wheeler-dealing was John D Ryan who had large investments in the Montana Power Co, the Great Falls Power Co., and the Chicago, Milwaukee & St Paul RR which had a part interest in the BA&P. The latter was itself about to embark on large-scale electrification. Electrified railways, using copper conductors and hydro-power from concerns in which he had an interest fitted logically into his plans.

The BA&P chose General Electric to mastermind its electrification. The 2.4kV conductor wires, suspended from cedar wood poles, were fed by two substations in which motor-generators converted 2.4kV ac into 2.4kV dc by means of two 1.2kV generators connected in series.

The orders came at a time when GE, the main advocate of direct current for traction purposes, was edging towards ever-higher voltages for main line work, now moving to 2.4kV, having tried 1.2kV on the Oakland suburban lines of the Southern Pacific (1911, the year in which it won the BA&P contract). The BA&P locomotives had four GE-229-A motors each, wound for 1.2kV and permanently paired in series.

The BA&P purchased 17 General Electric box-cab Bo-Bo locomotives weighing in at 80 tons each. By way of augmenting their power, the railway adopted the unusual ploy of coupling them, at times, to powered 4-wheel tractor trucks, 'calves' fed by electricity from the Bo-Bo 'cows'. The improvement in traction performance under the new electrified order was dramatic. Typically, main line traffic consisted of electrically hauled trains of up to 5,000 tons taking 1hr 45min for a journey, compared to 4,000 ton steam-hauled trains taking 2hr 15min, permitting a reduction of a quarter of trains along the main line, and with a good deal less locomotive preparation and servicing.

The reduction in locomotive costs was around 40%, and the electrification had paid for itself in five years. Many years later, at a Swindon Works debate, Charles Collett, the Great Western's Chief Mechanical Engineer, admitted that he saw the attractions of electrification, the main one being an escape from the thraldom of boilers. The BA&P returns throw light on this: no armatures failed in the first two years; motor commutators were not turned in 20 years; gear pinions lasted for 90,000 miles. In all, the cost of electric traction maintenance was around a third of that for steam[1].

The BA&P operations carried on for some 40 years, with a few minor additions to the motive power stock. Diesel electrics replaced the 1913 system in 1967. Even so, the line has its place in world history as the first high voltage dc main line electrification, proving that it could be done and paving the way directly for the extensive 3kV dc electrifications of later years.

Shortly after GE electrified the BA&P another contract came its way, from the Canadian Northern Railway (CNoR). In the years immediately before the Great War, railway building in Canada was proceeding furiously; two new transcontinental lines were being constructed to rival the Canadian Pacific: the Grand Trunk Pacific, and the Canadian Northern. The latter was a 10,000 mile assemblage of lines put together and extended by two visionary entrepreneurs, Sir William Mackenzie and Sir Donald Mann. Their Montreal terminus was, however, inadequate and poorly located. Accordingly they conjured up another vision: a 3-mile tunnel under the eponymous Mont Real (Mount Royal, 769ft, 234m) joining a prestigious new terminus near the waterfront to the main CNoR system, via a newly planned high-quality suburb, the 'model city' of Mount Royal.

The two-track tunnel was started in 1912, completed in 1918. Unfortunately, the conditions which had supported the recent building, arguably over-building, of Canada's railways went seriously problematic in late 1914. The steady flow of immigrants and capital dried up as the 1914-18 war took hold. The CNoR sank into insolvency and had to be taken over by the Canadian government, although the Mount Royal construction continued, albeit at a slowed pace.

The opening ceremony was distinctly downbeat; 21 October, 1918 – a train hauled by GE electric locomotive 601 bedecked with Union Jacks, then the national flag of Canada. The Great War was coming to its end; the Canadian authorities forbade public meetings on account of the Spanish Flu epidemic, and the glad, confident morn of the CNoR had long gone; no time for glee and squirting champagne.

Because of the great length of the tunnel, the new extension was electrified from the start, steam/electric engine changes being carried out at Lazard (now Bois Franc) about seven miles from the terminus. Following GE's advice, the CNoR adopted the modish, if short-lived 2.4kV dc system[2]. This was distributed by twin conductor wires connected

A 2.4kV dc Bo-Bo boxcab locomotive, as supplied by English Electric to the Montreal Harbour Commissioners, 1924-26. The GE boxcabs were very similar. Photo: *Former English Electric*

to a single substation containing two 1.5kW rotary converters (ac to dc) near the Eastern end of the tunnel.

GE supplied four hefty Bo-Bo box cab locomotives (Class Z-1-a), almost identical to the BA&P variety; a further two, with input from Canadian GE, were supplied later. This small fleet was augmented in time by some British-built locomotives, secondhand from the Montreal port authorities (Class Z-4-a, see below), and three steeple-cabs from GE in 1950 (Class Z-5-a).

After its quiet opening, the Mount Royal line settled down to a quiet and modest existence, at least, at first. So modest, indeed, that the usual moaners emerged advocating its abandonment, but wiser counsels overcame them. Long distance trains were hauled into the austere, minimalist Tunnel Terminal, later replaced in a massive urban regeneration scheme. A 9-mile extension was made to St Eustache-sur-le-Lac (now Grand-Moulin). Some useful commuter traffic developed, carried in unusual EMUs composed of converted, motorised steam stock cars sandwiching orthodox carriages. The converted stock (CNR Class EP-73) utilised GE motors, part of a batch originally ordered by the CNoR for EMUs from the Pressed Steel Car Co, but never constructed. As traffic grew, so the CNR ordered a much larger batch of cars (six motor, twelve trailer) in 1952.

Eventually the Mount Royal scheme fully justified the vision of its pioneers, but by then it was in other hands. It was converted to 25kV ac in 1994-95, given 58 new cars of various kinds now (2010) being augmented by Bombardier-built double-deck coaches. Ownership has also changed: the system is now operated by the AMT, Agence Metropolitaine de Transport. The last day of 2.4kV dc operations was 2 June 1995. Five of the six CNoR Bo-Bos, the original GE locomotives, have been preserved.

It had been intended originally that the electrified Canadian Northern line would link up with the network of port railways run by the Montreal Harbour Commission (MHC); in practice this was considerably delayed. Meanwhile, the Commissioners, attracted by the prospect of reducing pollution and standby costs as well as drawing on relatively cheap power, opted for electric traction in 1919 to power their system (19 route and 63 track miles). The MHC chose 2.4kV dc as their standard, to harmonise with the CNR Mount Royal system, which made two of its own locomotives available pending the procurement of the Commissioners' own motive power. In the event, they placed their order for locomotives not with GE but with English Electric.

The 100 ton Bo-Bo locomotives were accordingly constructed at the Dick, Kerr Works, Preston[3] on the electrical side, and on the mechanical by Beyer Peacock, Gorton, the first two delivered 1924 partly dismantled, aboard *SS Manchester Regiment*. Seven more followed up to 1926.

The EE locomotives had a similar appearance to the GE standard, but were better proportioned with longer frames, bogies, wheelbases, larger driving wheels and a narrow lookout proud of the main bodywork. Their dimensions were as follows (CNR GE type in brackets): wheel diameter, 4ft 5in (3ft 10 in); hp, 1,248 continuous (1,090hp); tractive effort on starting, 60,000lb (41,500lb) but maximum safe speed 35.5mph (50mph); weight 100 tons (83 tons).

These fine Lancasatrian products spent many years shunting and spotting trains, usually of grain from the prairies and sometimes weighing up to 5,240 tons; however, the Depression hit this trade hard and the Commissioners handed over their responsibilities to the Federal Government in 1932. In 1942 the system was de-electrified and the locomotives were swapped for some steam locomotives from CNR, becoming CNR Class Z-4-a. Thenceforth they took over a very different set of duties, hauling CNR commuter trains through Mount Royal tunnel. Four were scrapped in 1971, others later; one lasted until the end. None were preserved, unlike the more fortunate Mount Royal originals.

Americans occasionally refer to the third rail, the conductor rail of electric railways such as that which prevails in southern England, as the 'hot rail'. The hottest ever was born in the 2.4 era, a third rail supplying 2.4kV dc on the Michigan Railway in the USA, the highest third rail voltage employed on a railway or, more strictly, an 'Interurban' mainly concerned with passenger traffic.

The high voltage dc line ran some 90 miles from Kalamazoo to Grand Rapids and Battle Creek, Michigan and started operations in 1915. When entering built-up area it used 600V dc, overhead conductors, which also supplied the high voltage current at certain locations.

Much of the line, however, was third rail, fed by three substations, principally two which each housed two GE 1.2kV dc rotary converters connected in series. This audacious experiment did not last long, and the line was soon converted to 1.2kV dc.

Its faults were surely predictable: serious leakage of current, especially in damp weather; furious arcing in winter; arcing between the hot rail and axle boxes, only stopped by cutting off the current completely – and 'health and safety' concerns that confined waiting passengers to 'safety platforms' lest they wandered too near the deadly conductor rail which, surprisingly for such a touchy system, was an orthodox, bare top-contact third rail, admittedly with wooden board protection[4].

The Michigan experiment was the 2.4's only foray into third rail supply, except for some limited working in Chile (see below). Some of its rolling stock was remarkable: standard Interurban cars for stopping trains, but stylish, sharp-nosed, all-steel cars of great length (67ft 8in), built by J G Brill and the St Louis Car Co, for express services able, in theory, to touch 75mph, in practice rather more. Each mounted four, 150hp, 1.2kV GE motors; General Electric also supplied the braking and control mechanisms. This was, therefore, a railway firmly in the GE 2.4kV dc camp. For all its early promise, the line was soon in financial trouble; the former high voltage lines closed in 1928 and the remainder was abandoned the next year.

Mention was made earlier of some other 2.4kV lines, two of which were specialised mineral railways. In Italy, near Genoa, the FGC (Ferrovia Genova-Casella) is a light railway opened in 1929, operating on 2.4kV from the start and therefore rather a later entry into the group. Used by commuters and tourists, this 25km line climbs into spectacular terrain. This metre-gauge line later upped its voltage to the Italian standard of 3kV.

In Chile, the Bethlehem-Chile Iron Mines Railway (strictly, the FC Electrico de Cruz Grande a El Tofo) was built to carry exceptionally pure iron ore, 'way down South' in Chile for export to the USA from the Pacific port of Caleta Cruz. It was electrified in 1917 using the same GE system employed in the USA and Canada; part of the 2.4 'great tradition', such as it was. It ran three massive 120-ton locomotives as well as four 600V dc machines within the mining area. The 120-tonners had pantographs, but also third-rail pickups at either end of the line. The entire system was dismantled after 1973 when the concession ran out, even so the Bo-Bo locomotives had lasted an impressive 56 years under arduous conditions.

On the other hand, the lignite-carrying 2.4kV electric railway of the Lausitzer Braunkohle AG (LAUBAG) has employed a fleet of over eighty essentially German-designed Bo-Bo locomotives constructed at Henningsdorf in the 1960s, long and low machines of the traditional mining configuration quite common in Germany and once seen in the UK on the Harton Electric Railway. Some have recently been upgraded with chopper control, an arrangement quite unknown in the classic, GE era of the 2.4kV system.

This is no mean railway with its 240 miles of track, 56 level crossings, perhaps the apogee of the 2.4kV dc systems, but the least typical of them; untypical also in that the locomotives mount six pantographs, two orthodox, and four at the side for accessing current at loading points, and for remote control. This system and the C de F La Mure in France are the last remnants of the 2.4kV era, although neither is from the GE mainstream which sprung forth a century ago, a staging post to the more familiar 3kV dc, still very much with us.

The 2.4kV in our own era: a LAUBAG (Lausitzer Braunkohle AG) 2.4kV dc Bo-Bo on the Lusatian brown coalfield system in Germany, near Pietz on 9 May 2003 with two vertical and four horizontal pantographs. Photo courtesy *Michael Taylor.*

NOTES

1. Most of these statistics are to be found in Middleton, *When the Steam Railroads Electrified*.

2. As described earlier (Pages 7-10) Canada's only other main line electrification, the St Clair Tunnel (1908) of the rival Grand Trunk Railway, was a Westinghouse line; hence it employed the alternating current favoured by that firm, 3.3kV, 25 Hz. It has been suggested that the Canadian Northern warmed to the rival GE-dc system for that very reason. The St Clair system lasted 50 years.

3. The Dick, Kerr Works are cited as the electrical supplier by most authorities, although Middleton, op. cit., states Vulcan Works, Newton-le-Willows.

4. The Lancashire & Yorkshire Railway ran the UK's highest voltage third rail (Manchester Bury, 1.2kV dc) but was careful to arrange it in a well-protected side contact format.

REFERENCES

Charles V Mutschler, *Wired for Success*, 2002

Anthony Clegg, *The Mount Royal Tunnel*, 2008

J Edward Martin, *On a Streak of Lightning, Electric Railways in Canada*, 1994

William D Middleton, *When the Steam Railroads Electrified*, 1974

Joachim von Rohr, *The Tramways and Light Railways of Genova (Genoa)*,

Wilfrid Simms, *The Railways of Chile, Vol I, Northern Chile, La Calera-Copiapo*, 1999

Mark Walton, '*The Mount Royal Tunnel Electrification*' on the web, www.railways.incanada.net/candate/tunnel.htm

HENRY FORD'S DO-DO

A UNIQUE AND EXTINCT SPECIES

In full green livery, the DT&I No. 501 heads a freight train. In keeping with the American doctrine of high traction productivity, these trains could reach 1.5 miles in length.

Everything about Henry Ford's foray into railway ownership and electrification touched on the hyperbolic; even the unique locomotive he conceived presents some problems for wheel notations. It (they?) consisted of two double units, more or less permanently articulated in Do-Do+Do-Do. Americans, less troubled by taxonomical agonising, refer to the assemblage as a D-D type, or D-D+D-D if in its customary articulated arrangement.

This huge unit of motive power was No. 501 of the Detroit, Toledo & Ironton RR, a stupendous machine, weighing 375tons, claimed, not without reason, as the world's largest electric locomotive at the time, the mid-late 1920s.

The claim to uniqueness of No. 501 arose from a range of factors: it was never copied; nothing remotely like it was built before or after (although its electrical arrangements were to have a brief heyday as a model), and it must have been amongst the shortest-lived of all electric locomotives; it lasted 1925-31 and was then scrapped from whence it came, Ford's River Rouge motor works. The wheel arrangement was entirely without imitators, then or ever. But before turning to its unusual and, as it turned out, successful electrical engineering, why No. 501 in the first place?

Henry Ford was himself a unique type, astonishingly creative and original, but also apt to adopt unexpected, occasionally wrong-headed, views and initiatives. Increasingly fed up with the service and cost of transporting his motor vehicle products from Detroit to elsewhere in the USA, or for export, by the established railway system, he decided to join them by acquiring a line of his own, extending and reorganising it, and thereby administering a hard kick to the established system.

The cost of the standard Model T Ford (about $260 and falling, at the time; 10,000 were being built each day) was famously quoted as 'f.o.b. Detroit', i.e. 'free on board' where loaded. The purchaser, often a dealer, then had to pay for 'c.i.f', carriage, insurance and freight. Obviously, if Ford could reduce the initial c.i.f by means of his in-house railroad, it would add to the price attractiveness of one of his cars, tractors or trucks.

Ford's chosen instrument for this policy was the 500-mile Detroit, Toledo & Ironton RR (DT&I) which connected 'Motor City' (as it then was) to other systems. It was entirely steam operated at the time of its acquisition (1920), and becoming distinctly run-down, referred to by wits as the 'Darned Tired and Irksome'. Ford was soon at work transforming

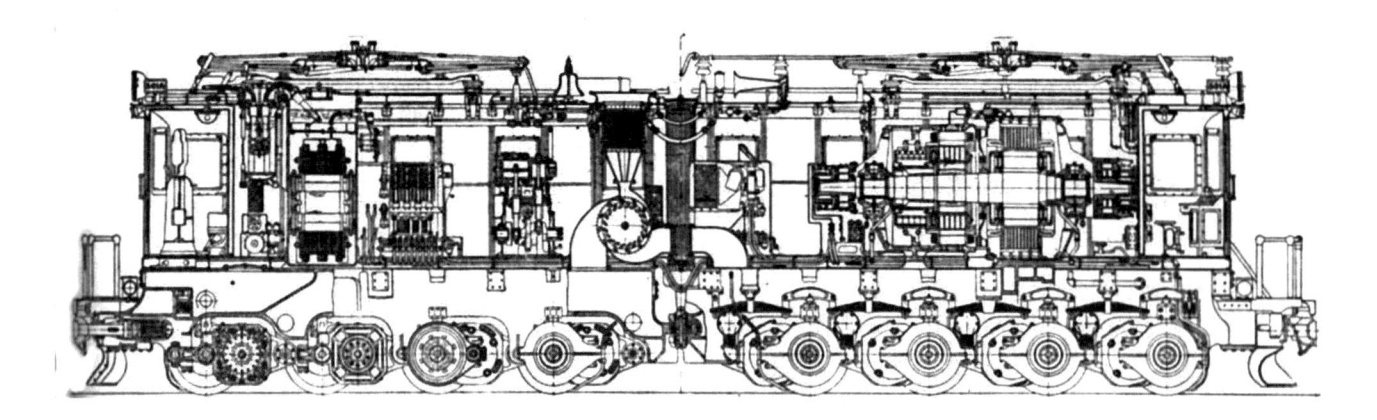

Above: Cross-sectional elevation of a single Do-Do unit. The massive motor-converter was housed in the right-hand half.

Below: Single DT&I No. 501 when new, reposing beneath the high voltage ac catenary, held aloft by Ford's concrete gantries of great solidity

the DT&I radically. First, the steam fleet was refurbished and upgraded to high standards. The impeccably maintained locomotives were given nickel-plated instruments and cylinder covers, buffed coupling rods, upholstered aluminium chairs for the crew, and lacquered boiler jackets.

Since the DT&I was principally a freight and mineral hauler, its typical motive power was composed of 2-8-0s and some 2-10-0 'Decapods' originally intended for Tsarist Russia, but held back in the USA because of the Bolshevik Revolution. Passenger traffic was in the hands of some 4-6-0 'ten wheelers', but true to form, Ford went against the grain and purchased two 4-4-2 'Atlantics' from the Michigan Central, an obsolescent form, but high-stepping and elegant, all the more so after they got the full lacquer and nickel treatment. Fortunately, one has been preserved.

Footplate crews, like all DT&I employees, had their wages increased and generous welfare facilities offered - on the other hand, Ford's oppressive paternalism soon manifested itself: no smoking on railway premises; crews to be clean shaven, overalls to be immaculate. Some of the permanent way, bridge repair and breakdown crews were treated to upgraded accommodation cars, complete with kitchens and three meals a day, from bacon and eggs at breakfast to 'steak and two veg' in the evening. However, in accordance with his brisk, no-nonsense style, Ford cleared out the ranks of management on the grounds that 'fewer and better' would do the job.

Ford's penchant for radical reform led him to initiate a 17-mile cut-off, the first step in his plans for main line electrification. The infrastructure for electrification, completed in 1926 was without peer, so thoroughly contrived and constructed that some of it has endured to this day; long after the electrification itself was dismantled.

In common with his general upgrading policy, Ford saw to it that permanent way standards were set high; well-husbanded track, regularly inspected. He adopted concrete in a big way, rather tending to overdo it. His concrete sleepers were a good idea, but when laid on unyielding concrete blocks they had a tendency to shear. Far more spectacular were the long stretches of massive concrete gantries from which the conductor wires were to be suspended.

Although we associate Henry Ford with motor cars, trucks, tractors, etc., he was quite at home in the field of electrical engineering, having once worked with no less than Thomas Edison, and having been Chief Engineer of an electrical undertaking, Detroit Edison. Herein lay the source of his most radical idea for the DT&I - its eventual electrification, possibly linking it to the Ohio River, even to the already-electrified Virginian Railway far to the South, thereby giving his Detroit motor works access to the Atlantic seaboard.

Electric power was to be generated at Ford's River Rouge plant in Dearborn, with three new turbo-alternators set aside for powering the railway. These produced three-phase ac 13.6kV at 60Hz, fed to a substation that converted it to single-phase 3.6kV ac, 25Hz for the railroad, stepped up variously to 11, 24 or 46kV. This current was usually fed directly to the conductor wires at 11kV, although later in the life of the great experiment this was stepped up to 22kV.

At this point the gantries came into play, massive reinforced concrete arches chosen because of their long term ease of maintenance, ironic given the relatively short life of the electrified DT&I. Altogether 365 of these arches were cast at the River Rouge plant, each consisting of 2-5 separate but standardised components depending on the precise location and function of individual supports. The normal arch weighed over 30 tons - hugely over-engineered for the bronze conductor wire that carried the high voltage current. Many were left after de-electrification, the cost and hassle of dismantling such architectonic solidity was hardly worth the labour and expenditure involved.

The electrified DT&I was launched with a typical Ford PR onslaught. He had long realised the value of strong publicity, and was himself no mean phrase-monger; his views are often quoted to this day on, for example, the choice of colour for new cars, or the nature of historical study (see later). Having already outlined his general electrification strategy in 1923, on 29 November 1924, the intended 'world's largest electric locomotive' was announced by Henry Ford. He outlined its many advantages, drawing attention inter alia to its ease of control, rugged construction, ability to regenerate power on downgrades, or when decelerating - this would be awarded the laurels of being 'eco-friendly' today.

In the event, only a 16-mile segment of the railway was electrified (River Rouge complex - Flat Rock freight yard, Michigan), and there was to be but the one electric locomotive working it, but what a locomotive! There were arguably two or four of them, cast in the form of two, twin locomotive units more or less permanently coupled as a Do-Do+Do-Do form, although on rare occasions, as noted, a Do-Do unit operated singly. Livery for the ensemble was dark green with red highlighting on the wheels, a striking combination and, be it noted, not standard Model T black.

This locomotive of locomotives was to be the 375 ton DT&I No. 501 (i.e., 501A and 501B, linked). Single-phase ac was stepped down as necessary in each twin unit to 1,250kV; one part unit bore transformers, switchgear, blowers, etc., whilst the other housed the 29 ton, 750rpm motor-generator, turning the ac into 600V dc for the eight nose-suspended traction motors in each twin unit. The outcome, which could drain the power station on occasion, was a maximum tractive effort of 250,000lb, or 3,400hp continuous rating. Normal speed was 17mph, with motors in series; top speed was 25mph, motors in full parallel with weak field. Control was precise: 45 steps between 0 and 17mph by regulating main generator voltage; and 17-25mph by varying the excitation of the axle-hung traction motors, final drive being through flexible quills. The motor-generator was spurred to action by drawing on dc batteries, later re-charged as the entire assembly came to life.

Frames, superstructure and motors came from Ford's own works; Westinghouse, the ac specialists, looked after electrical apparatus, hence the usual designation, a 'Ford-Westinghouse' locomotive. With typical Ford originality, all bearings were inside; leading and trailing axles had 2in lateral motion, other axles were individually braked.

The DT&I worked its electrical giant hard, '24/6' in that Henry Ford had strong views on Sunday working which was forbidden, except for emergencies. In their glory years, 1927-31, the twin electrics were running 34,000 miles per month. They were, however, gargantuan consumers of electricity, and cost $2.28 per mile to run, against $0.93 for steam traction, although electrical maintenance costs were much lower than steam.

Even before Ford sold the DT&I, its heavy costs, unrealised extension plans, insufficient density of traffic, and onset of the Depression spelt the end of this noble experiment. He sold the DT&I to a Pennsylvania RR subsidiary in June 1929. The electric project ended; scrapped in 1930, another Do-Do became extinct. The wires were unstrung, although many concrete gantries remained to remind the world of that which might have been.

One outstanding feature of Henry Ford's Do-Do system was its attribute, effectively, as a mobile ac/dc conversion substation in a world where this conversion was almost universally carried out by sending high voltage ac to fixed substations built beside railway track, as for example on the UK's Southern Railway that required such constructions virtually every 2-4 miles.

The development of current and frequency conversion systems on railways has been an international enterprise, unlike the invention and early development of the steam locomotive, which, in an era of slowly spreading technical knowledge, was an overwhelmingly UK, even Northumbrian initiative.

Above: A double rarity: a 'half 501' at work, hauling a passenger train on a special occasion, various Ford engineers and executives decked out in straw 'boater' hats, fashionable at the time. Two variants of the concrete gantry evident, for single and double tracks.

Below: Henry Ford with his classic Model T car, a staple traffic for the electrified DT&I.

The problem of how best to take advantage of the ease and economy of ac distribution, and flexible, convenient dc traction systems exercised many scientific and technical minds in the early twentieth century; it was in this context that Ford's experiments took place.

As early as 1905, General Electric outlined a possible answer for the New York Central RR, a normal dc locomotive (1-Bo-Bo-1) hauling a massive trailer that could pick up and convert alternating current. Nothing came of it, although GE did attempt a lash-up, with some success, in 1910. One problem facing the engineers was a conundrum: mechanical conversion was entirely feasible, but involved great dead-weight, motor-generators that weighed heavily upon the earth, acceptable in a fixed substation, but less so on moving trains where axle weight had to be minimised. On the other hand, the early mercury-arc rectifiers that could more conveniently change ac to dc were lighter but also delicate, capricious, and unreliable.

Of the 18,000 or so miles of electric interurban railways in the USA, only one tried a mobile rectifier, Paul Smith's Railway, a short line serving a luxury hotel in the Adirondacks. Even there it proved problematic, and was replaced by a motor-generator. Both General Electric and its principal rival, Westinghouse, experimented with mobile mercury-arc rectifiers, 1911-12, concluding that they lacked reliability. Meanwhile, in France, the PLM railway developed a successful ac/dc generator locomotive by using mechanical conversion, a 2-Bo-Bo-2 of 1912.

Henry Ford's Do-Do was an example, therefore, of the playing-safe school, carrying out conversion by rugged mechanical means. Its immediate background was the successful attempt by the Norfolk & Western Railway to convert single-phase ac to three-phase ac on board locomotives, starting in 1915. If ac to ac, why not ac to dc as the PLM had already demonstrated to be feasible?

The principle of motor-generator locomotives was to have but a brief heyday, however. It was adopted by the Great Northern Railway in the USA (1927) and later by the New York, New Haven & Hartford RR and the Virginian Railway. But their massively heavy technology was eventually overtaken by practical mercury-arc rectifiers, semi-conductors, and later the more sophisticated frequency generators of our own era. These take either ac or dc and invert it to three-phase ac at variable frequencies, another saga altogether. For all that, Henry Ford was on the right track, trying to combine the advantages of ac distribution and dc traction control, exactly like the well-remembered first generation of British Railways ac electrics.

Afterthought: is history 'bunk'? A brief note, in conclusion, on the subject of Henry Ford and history. Whilst it is true that he once stated, under extreme provocation during Court proceedings that History was 'more or less bunk', the context in which he made this assertion suggests fairly clearly that he was referring to the 'kings, queens and battles' type history. Later, he made generous donations by way of setting up themed museums and a heritage centre at Greenfield Village, to demonstrate 'history as it really was'. To the Great Mechanical, that meant history as determined by science and technology. The jury is still out on that. His other pronouncement that 'The only history worth a tinker's damn is the history we make today' suggests that fascinating although he found some aspects of the subject, a rather inflexible and aggressive pragmatism may have coloured his view of its study, as it did of much else in his remarkable life, his great Do-Do assemblage included.

EDITOR'S NOTE.

1. This chapter first appeared as an article in *Locomotives International* No. 87 and is reproduced here by courtesy of its Editor.

THE BALTIC ELECTRICS

No.13: the well-known pot-boiler photograph of the Merz-Raven Baltic in shop grey, but demonstrating its good looks – and quill drive – most clearly. Why 13, of all numbers? Because the NER had a dozen electrics already. Because the NER declined to use the 'E' suffix employed by many electric operators, a great deal of number-shuffling had to take place; the original No.13, an F8 2-4-2T became 40. The problems we make for ourselves. Photo: *Former AEI-EE Traction*

Electric locomotive evolution, over a century old by now, seems to have stabilized, outwardly at least, with the near-universal double bogie type: Bo-Bo, or Co-Co. Inwardly, however, there are many variations regarding, for example, voltages, frequencies, forms of control or current conversion.

In the 1900-20 period an enquiring traveller might come across a wide range of electrics: nose suspended motors within bogies; 'gearless' motors, with armatures on axles; rod-drive locomotives, looking somewhat like steam engines 'below the belt'; and a further set of variations whereby motors mounted in the body of the locomotive powered large driving wheels by various kinds of sprung or adjustable linkages, such as the 'quill-drive', and the 'Buchli drive' that was favoured in Switzerland.

The German-originated 'VDEV' wheel classification was adopted early for electric locomotives; it became virtually the world standard in the 1930s as the International Union of Railways 'UIC' classification[1].

Earlier, it was fairly common to employ Whyte, steam-based forms for electrics. The LNER generally resisted UIC usage to the end. Thus, its Shildon-Newport Bo-Bo locomotives were usually described as 0-4-4-0 electrics. The unique NER electric express locomotive, No.13, constructed for 'the electric railway that never was'[2], the York-Newcastle scheme of 1919 was a '4-6-4' in LNER parlance.

Since a 4-6-4 was a 'Baltic' locomotive in the world of steam, the phrase might also describe this unusual form of electric, although it seems rarely to have been employed. Even so, 'Baltic electric' is a neat way of summarizing plenty of evolutionary development in a short space[3].

NER No.13 The story of Vincent Raven's ill-fated 2-Co-2 electric locomotive is fairly familiar, and has been often told[4], a bare outline must suffice here, but with some additional contextual material.

After the Great War, faced with growing pressure of traffic on its York-Newcastle main line, now part of the famed 'ECML', and with a good record of electrification to its credit, the NER investigated the possibility of electrifying some 80 miles of main line, with adjuncts. The Chief Mechanical Engineer, Sir Vincent Raven strongly supported the idea, and the NER's main electrical consultants, Merz and McLellan, thought the idea to be entirely feasible.

The post-Great War boom tailed off rather suddenly, however. The new Ministry of Transport, flexing its muscles (or throwing its weight about) requested that the NER put matters on hold whilst the Kennedy Committee advised on national standards for railway electrification, particularly with regard to voltages. The tide had been missed; the directors' feet went cool, especially with the 1923 Grouping imminent. Thereafter the scheme

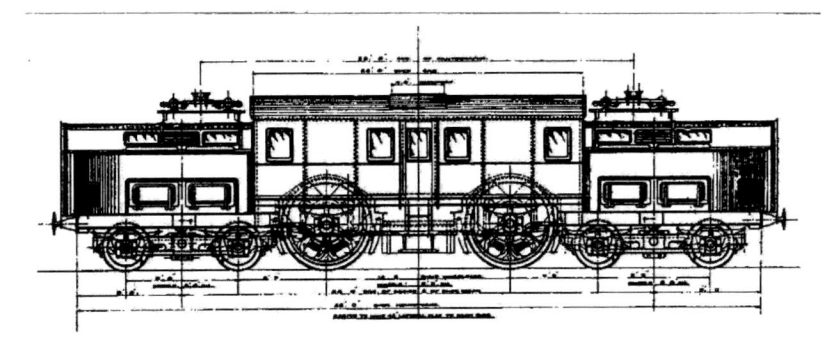

NER electrics, designs drawn up by Merz & McLellan for the abortive York-Newcastle scheme of 1919; closely following Westinghouse quill-drive practice. If PRR experience was anything to go by, the 4-4-4 (2-Bo-2) would have had adhesion problems, like the steam versions on the NER and Metropolitan railways.

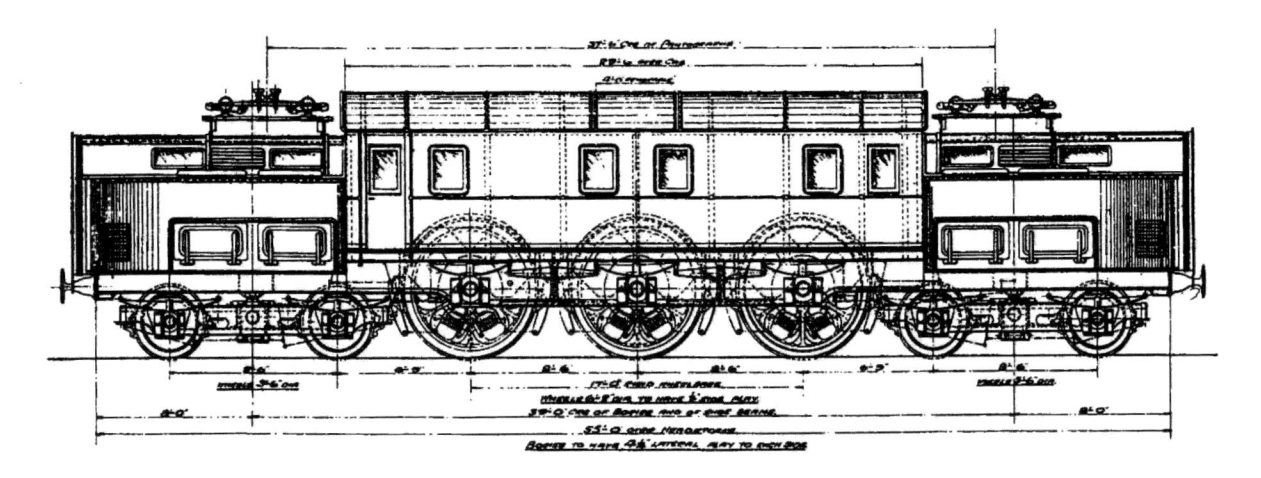

lay in the shallows, in the land of might-have-been. Seventy years later, the electrification came under the aegis of BR, but that is quite another story.

In 1919, America led the pack with regard to main line electrification and had not yet handed on the baton to Europe. Its achievements included some eighteen thousand miles of electric 'Interurbans'; street tramways; elevated urban lines; great tunnel electrifications (Hoosac, St Clair; Cascade; Baltimore; New York) and perhaps most impressive of all, the first long-distance main line electrification. This was of the Chicago, Milwaukee & St Paul RR in the mountains of the West; eventually over 600 miles of 3kV dc railway – the leading edge of electrical technology, or so it seemed.

Raven, and Francis Lydall, of Merz and McLellan, beat the predictable path to the USA; Raven's report to the NER board on the feasibility of main line electrification was saturated with American examples.

For its first wave of electric locomotives, the 'Milwaukee Road' offered a good object-lesson: it ran locomotives with gear-driven axles as well as the far-famed EP-2 (electric passenger type 2) class with their 'bipolar, gearless' drive, with armatures mounted on twelve driving axles.

Raven warmed to the 'bipolar' idea, and later tried to get the NER directors to approve the building of a 2-Co+Co-2 example at Darlington. It was, however, the Milwaukee's very latest acquisition that impressed Raven and Lydall the most. Unlike the earlier machines, this was from the Baldwin-Westinghouse stable; the class EP-3 'Westinghouse Motors', in railroaders' vernacular: 'the quills', or occasionally something slightly stronger when their inadequacies came home.

These impressive looking 2-Co-1+1-Co-2 electrics ran to 88ft long and 17ft high: Russian proportions! Their 5ft 8in driving wheels bore the quill drive that Raven tried out on NER No.13. This was a Westinghouse speciality, having been already pioneered with fair success on the 'New Haven': the New York, New Haven & Hartford RR, parts of which had been electrified 1907 onwards; a pioneer in the use of ac traction, 11kV, 6.6Hz.

The New Haven's small fleet of 1-Bo+Bo-1 electrics had demonstrated the virtues of quill drive. This oddly-named 'quill' was, in fact, a hollow tube surrounding the axle, with typically 1.5-2in clearance. The motors drove this tube which engaged one, or more usually both of its ends, with the actual driving wheels, via a multi-armed 'spider' and strong, horizontal, helical springs. The idea was that the axle could rise and fall within the quill, thus coping with the shocks of uneven track. Also, by mounting motors on the main frame, unsprung weight was minimized.

The New Haven liked its quill drives, and they became standard for its traction. The Milwaukee version was, however, a pain. It led to acrimonious correspondence and fraught meetings between the railway and manufacturers, Baldwin and Westinghouse. Although the EP-3's rode smoothly and offered an even torque, they soon became a fitter's nightmare as tyres wore out swiftly, quills cracked, and springs disintegrated at speed. Baldwin had underestimated the necessary strength of the frames, which were prone to crack. Various palliatives were tried over the years, but the EP-3's remained problematic.

Would No.13 have disappointed in this way also? Probably not: the Milwaukee Road was a

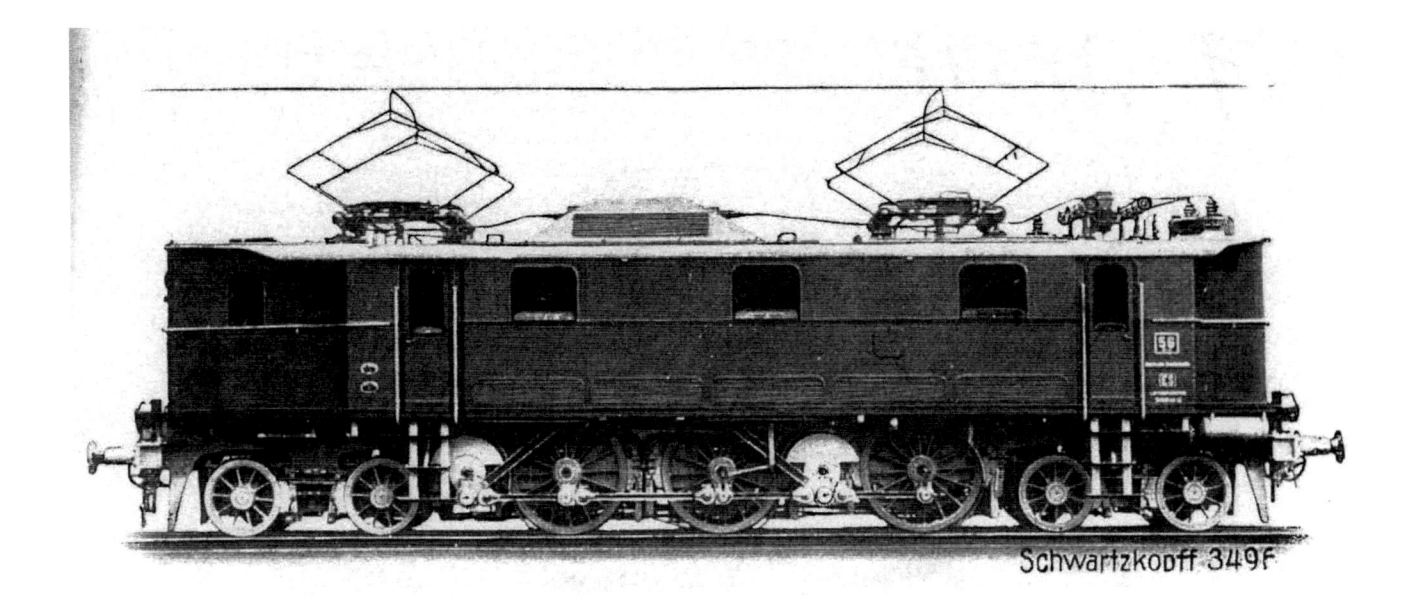

Above: Rod-drive swansong, the last of the German rod-drive classes, for express work; the only 2-D-2 configuration. The prototype form, designed in Prussian days and with cabs set back;

Above right: Stretched Baltic: one of the experimental 2-Co-2 locomotives tried out by the Great Indian Peninsula Railway; Hawthorn Leslie and Brown Boveri, 1929. The driving wheels of this 2,160 hp machine are masked by the Büchli drive shields. As it turned out, the 'electric Pacifics' were preferred, with their greater proportion of weight for adhesion. This was GIPR class EC/1; IR class WCP-4.

Below: The later form, as E06 12 fresh from Schwartzkopff in 1925: cab set forward. The boxiness of the earlier type was offset by some 'form follows function' asymmetry, to their aesthetic advantage. Schwartzkopff (Berliner Maschinenbau)

line of sinuous curves, heavy gradients, seriously inclement weather, and demandingly heavy trains. At least, that seems to have been the Westinghouse reaction, mindful that the New Haven quills were a pronounced success – but they ran on level track, often straight and where not, easily curved and cambered. Since the York-Newcastle equates more readily with the New Haven's line from New York to Stamford, Raven's judgment was probably right.

In spite of a persistent legend that No.13 had large driving wheels because Raven was a steam engineer with a related mindset, their true origin lies elsewhere, mainly in Westinghouse's drawing office[5]. Early electrics often had relatively large, heavy electric motors, better mounted on main frames rather than within track-punishing bogies. The development of electrical technology was to render these solutions obsolescent, even as the Baltics rolled out of their constructors' shop.

There is a curious mystery about No.13 (not its number, which was logical in the NER electric sequence, even if asking for trouble in the eyes of the superstitious) – its specifications, sent to Metrovick who supplied its electrics, included the ability to start a 400-ton train on 1 in 78. No such gradient afflicts the ECML between York and Newcastle, but there is one such between St Margaret's and Waverley, Edinburgh. So, perhaps Raven's expansive talk about electrifying 'to Carlisle and Edinburgh' was the outward sound of plans as yet unrevealed[6].

No further Baltic electrics were tried in Britain, although two more were to be built there. As for No.13, by the time it was scrapped in 1950 it had one of the lowest mileages of any electric locomotive anywhere, and probably a world record for not having operated commercially over its 35-year life, although there is an enduring story that it returned some empties from Newport to Shildon on occasion, a kind of commercial operation.

We shall never know how a fleet of NER electric Baltics might have turned out. No.13 had some teething troubles, such as overloaded contactors, and some signs that overheated motors might have resulted if they were run flat out for long periods – but these typical breaking-in problems were of a soluble kind.

Although NER No.13 was not to be the harbinger of a British electric fleet, in contemporary Germany the 2-C-2, the rod-drive Baltic electric, nearly achieved fleet status. It was a close-run thing, and a dozen were built before the new Deutsches Reichbahn, DRB decided to abandon rod drive for its electrification plans in 1925.

The German saga started in the post-war twilight of the former Prussian State Railways (KPEV) in 1919, at the point they merged into the DRB, the state railway company of the new order, the 'Weimar Republic'. The Prussians were leaders in the European surge towards long-distance electrification that was to put American work into the shade, largely because the Europeans could tap massive public funding for their schemes. The USA, firmly wedded to private railway ownership, was unwilling to do this.

In common with other German schemes (as well as those in Austria and Switzerland) the KPEV chose alternating current for its railway electrification; this had more technical potential at the time than, for example, the Milwaukee's high voltage dc. The Prussian standard was ac at 15kV, 16.66 Hz.

In late 1919 the KPEV took delivery, from Schwartzkopff, of five massive 4-6-4 electrics, numbered ES51-55. Their numbers indicate their purpose, ES being Elektrische Schnellzug, electric express engine. Their top speed, 110km/h (nearly 69mph) was respectable for their duties: fast passenger trains in the Halle and Hanover areas, each being a Reichsbahndirektion, a distinct administrative region.

The ES series, perceived as a possible standard for the future, was to be the electrical equivalent of the Prussian class S10/1, an express steam 4-6-0. The configuration of the locomotives drew on experience with some successful Baltic tank engines, the Prussian T18 class, and the Alsace-Lorraine (Elsass-Lothringen) T17 class, both 4-6-4T types, of a kind popular about the time of the Great War.

Above: The slippery class O1; 01c 7856 in 1931, the 4-4-4 (2-Bo-2) that was meant to replace the legendary class E6s Atlantics, they suffered ultra-Atlantic wheel slippage; Baltics were to be preferred. Photo: *Erstwhile PRR.*

Each ES locomotive bore a massive, frame-mounted ac motor. This powered the 1,600mm (5ft 2in) driving wheels via cranks, placed between a bogie and the first driving axle, and between the next set of driving wheels. The whole arrangement was set in a ponderous, circling triangle of driving and connecting rods.

Two more of the ES locomotives were added, ES56 and 57, distinguished from the first batch by having air-cooled transformers instead of the oil-cooled variety of the originals. By this time the KPEV was merged into the DRB; the class was given a formal designation (class E 06) and new numbers were awarded, E 06 01 to 07.

The Baltics were promising, able to meet their specification of taking 600-ton expresses, or 500-ton slower trains along the fairly flat terrain for which they were intended.

Soon afterwards, the DRB considered a range of standard locomotives (steam as well as electric); the 2-C-2 type was one of the original favourites. The initiator of the standard electric proposals (1922), Dr-Ing W Wechmann, drew up seven categories of locomotive. The first two were for express locomotives in (i) Flachland, level terrain – 2'C2' and (ii) für Schnellzüge in Hügelland, express trains in hilly terrain, 1'Do1'. The precise (pedantic?) German usage employs an apostrophe to denote a movable, tracking bogie, truck, or axle. The VDEV notation for these locomotives, in its full glory, was: 2'C2'-w1k-S; meaning: 4-6-4; w - Wechselstrom (ac); 1 – one motor; k – Kurbelantrieb ohne Vorgelege, crank-driven without reduction gears.

The E 06 designs were dusted down, and five further, improved Baltics were constructed as class E 06/1, duly numbered E 06 08 to 12. They were constructed by Berliner Maschinenbau, electrical equipment by Bergmann. The chief alteration was to give the new locomotives flat-fronted forward cabs at each end; the originals had cabs set back with the driver's window looking out along the side of the engine. Forward of these early cabs was sited equipment, for example a train heating boiler.

The E 06/1 type was set to work between Leipzig and Magdeburg, via Dessau. In 1934, further electrification made possible a circular service starting and finishing at Leipzig, going via Magdeburg and Halle.

The German Baltics nearly made it – but not quite. They were launched in Prussia, just as that jurisdiction was merged into the new Germany, and then tried out as a possible standard type, at the very point the DR decided to abandon rod-drive for its larger and faster electrics.

Why this decision? Largely because the rod drive configuration (Stangenantrieb) arose from the early preference, even necessity, for having one or two huge electric motors; equipment of such weight that it was necessary to mount them on the main frame. There was a price to pay for this: rod drive was generally more expensive to maintain than its rivals, and it replicated the typical steam traction problems of requiring counterbalances, tending towards unsteadiness and hammer-blows at higher speeds.

On the other hand, as Francis Lydall pointed out to the Institution of Locomotive Engineers (1949),

Above: One of the standard P5a Baltic electrics of the PRR, not too bad an example of the shoebox school of design, helped by 6ft driving wheels, but hindered by the sensible, albeit conspicuous and sinuous safety rails, front and rear. Built by General Electric, 1932; withdrawn 1963. Adhesion problems hinted at by the generosity of sandpipes. Seen at Potomac Yard, Va in February, 1946. Photo: *Harold K Vollrath*

Below: Last of the Baltics: Pennsylvania class P5a Modified, 4746. Technically similar to, but more potent than the standard P5a type, and given the 1930s streamline treatment, refined by stylist Raymond Loewy in the image of the more famous GG-1 electrics. The cab was set back mainly on 'health and safety' grounds; the crews had poorer visibility but a whole lot more locomotive in front of them in the event of a collision. A handsome machine, used for passenger and freight work; Washington DC, May 1938. Photo: *Harold K Vollrath*

Quill-drive: the Westinghouse-based drive that powered most of the Baltic electrics, adopted in Northumbria, India and the USA. Here (at Darlington North Road works) reposes one of NER No.13's driving wheels, a Vickers product. The axle-surrounding quill, ending in a many-armed 'spider' will press on the main spoke trunks through large, horizontal, helical springs – hence the trifurcation of the spokes by way of easing out the resulting stress. At 6ft 8in, these were the largest driving wheels on any Baltic electric, possibly on any electric locomotive.

rod drive resulted 'in a higher coefficient of adhesion than for individual axle drive'. He cited in support an embarrassing incident when NER No.13 slipped helplessly at Shildon one day, having to be rescued by a 4-6-0, much to the annoyance of Sir Vincent Raven. Neater, lighter motors had come on stream by 1925; the great years of rod-drive were over.

The twelve E 06's nevertheless played out their lives usefully, the first spies of battalions that never came. They had all gone by 1956.

* * *

Two more Baltic electrics were constructed in the UK in 1928 for India, experimental prototypes of classes that were not to be. The mind of Charles Merz can be detected behind them since Merz & McLellan were the consulting electrical engineers to the Great Indian Peninsula Railway that commissioned these unusual machines. The GIP was an early entrant into electrification in Asia, partly to ease the pressure on its Mumbai (Bombay) suburban services, partly to relieve the operating bottleneck presented by the Ghats, with gradients of up to 1 in 37.

The most numerous and successful of the GIP electric locomotives (all were designed for 1,500V dc) were rod-drive 'hunks', ideal for slow freight, banking, etc on the Ghat gradients. This was class EF/1 (Indian Railways class WC1G), forty-one C-C engines with input from Metrovick, SLM, and Vulcan Foundry.

Two Baltic prototypes were tested for passenger work. They were separately classified: EB/1 (IR, WCP3), and EC/1 (IR, WCP4). Each was a platform for evaluating a different form of drive. Thus, EB/1 (Hawthorn, Leslie; GE) contained a novel form of quill-and-yoke drive, based on Oerlikon practice that might be best described as 'not uncomplicated'. The device was explained neatly in the *Locomotive Magazine*, 15 August, 1928:

> 'The drive… transmits torque with varying eccentricities between centre line of axle and centre line of quill without rapid acceleration or deceleration of the armature masses…'

The EC/1 used the more familiar Büchli drive, a Swiss device that was also complex, but proved to be reasonably reliable over the years. In this case, the motors (two per axle) drove gears within a shield that could rise and fall slightly, relative to the axle. The shields masked the driving wheels in this Anglo-Swiss offspring of Hawthorn, Leslie, and Brown Boveri.

The EB/1 looked rather neater, although both locomotives had a rather stretched-out appearance if viewed broadside on. Merz & McLellan drew up the specifications which included an ability to pass through floods 2ft 9in above rail level; a tractive effort of 24,000lb up to 36mph, and 6,300lb at 70mph; a maximum speed of 85mph, and the ability to start a 450 ton train on a 1 in 100 gradient. This feat had to be accomplished ten times, at five minute intervals, without the temperature of the main resistances exceeding 250 deg C.

As with all Baltic electrics, the motors were mounted on the main frames, and were therefore part of the sprung weight. Merz suggested three prototypes, the two Baltics, and two 'electric Pacifics', 2-Co-1. In the event, the Pacifics won favour and a contract for twenty-one of them was placed with SLM and Metrovick.

The 'Pacifics' were more compact (49ft 4in as opposed to 54ft) with a slightly higher proportion of their weight resting on driving wheels. Sixty out of their hundred tons weight was available for adhesion; the Baltics weighed about ten tons more. The drive of the 2-Co-1s was similar to that of the Baltics; one used the flexible, internal drive, the other employed the Swiss external drive. The two-wheel pony truck of the Pacifics was linked to its neighbouring driving axle 'so as to form virtually a four wheel truck' (*Locomotive Magazine* 15 Feb, 1929), on the Zara principle familiar in Italian steam traction. One of these machines is in the collection of the Indian Rail Transport Museum.

The Baltic electric form had chalked up another near-miss and that might have been the end of their saga. As it happened, their greatest hour was just about to dawn.

In the late 1920s and early '30s the Pennsylvania Railroad took over the leadership in American railway electrification from the increasingly cash-strapped Milwaukee Road. The Depression hit the Pennsy's plans hard and they only survived by dint of Federal government assistance[7]. The major PRR scheme was electrification from New York to, variously, Philadelphia, Harrisburg and Washington dc. The traction current chosen was 11kV ac.

Although most railway cognoscenti would probably associate the fine, streamlined class GG1 (2-Co+Co-2) locomotives for this great initiative, the first wave of motive power chosen was, improbably perhaps, a fleet of Baltic electrics.

The PRR worked closely with Westinghouse on traction matters. The candidate locomotives for the new network emit ghostly echoes of the NER York-Newcastle proposals a decade earlier. The plans drawn up by Raven and Merz contained both 4-4-4 (2-Bo-2) and 4-6-4 (2-Co-2) varieties for light and heavy passenger traffic respectively: precisely the same as suggestions entertained by the Pennsylvania. In this case, however, both prototypes were constructed, and one went into large-scale production.

At first glance it might seem that having so much non-adhesive 'wheelage' was unwise. But that would be to ignore the benefits accruing to the PRR from having splendid, strong track, ways and works. Its E6s Atlantics, for example, had around 30 tons bearing on each driving axle, mitigating the 4-4-2's tendency to wheelslip; 'losing traction' in later argot.

Similarly, the axle loadings of the prototypes reached levels unobtainable in most railways, anywhere; the same eye-opening 30 tons. This permitted loading up to 6,400 hp (one hour rating) on to the driving wheels, or 2,150 hp per axle, and a starting tractive effort of 55,000 lb; not, as it turned out, too wise a move[8].

The 4-4-4, class O1, was nevertheless a notorious slipper, like its equally elegant steam cousins of that wheel notation. There were ideas abroad of using it to replace the E6s Atlantics; in the event only eight were built.

Unusually for a railroad that adopted new classes only after thorough investigation and evaluation, the PRR took a plunge with its Baltic electrics in 1931, ordering 90 new ones, straight off; class P9a, having briefly tested two prototypes, class P9. It paid a heavy price for this audacity. The P9a Baltics soon proved to be trouble-prone as 1933 wore on. At one point they had to be withdrawn en masse, giving steam supporters a good chortle when existing duties were undertaken by K4 Pacifics.

The P9a's suffered three main initial problems: they swayed about unsteadily at speed; they never quite matched hopes and expectations of shifting fast passenger trains – and most seriously of all, axles started to crack. The unsteadiness was alleviated by redistributing weight and improving bogie design. The axle problem was attributed to having too narrow an axle for so high a torque, and to indulging in the false economy of having quill-drive on one side only. Thicker axles and twin quills were installed.

The slightly disappointing performance was never really surmounted, although the class was generally reckoned to be a success and it was long-lived. For future long distance work, the Pennsylvania commissioned the immortal GG-1 class of 2-Co+Co-2 engines.

There was a coda to all of this: the last batch of P9a's was redesigned and made more powerful. Crews were given better protection by moving cabs back along the frame. The result of all this was the 'P9a Modified' class, modelled like the GG-1 class on a streamlined form finessed by Raymond Loewy.

These last Baltic electrics of them all had striking looks; the saga ended as it had begun with NER No.13, with quill-drive machines of impressive appearance. Unlike the rather stretched 4-6-4 electrics elsewhere, No.13 and the P6a Modifieds approached more nearly the 'Golden Ratio' of the Greeks, 1:1.6; like the Atlantics, they were well-proportioned.

These, then, were the main line Baltic electrics; a rare breed but one that encapsulates a great deal in an era of rapid transition. There was also another, the rarest of birds: a diesel-electric pseudo-Baltic, built for the NWR in India, but actually of the odd configuration 1-A-Co-2; i.e. one bogie was part-powered. It was constructed by Armstrong-Whitworth in 1935, and given a Sulzer diesel motor.

The 4-6-4 electrics proved to be branches off the main trunk of evolution that was leading inexorably to the all-adhesion Bo-Bo and Co-Co types with which we are now familiar. They taught lessons that had to be learned, and were generally reliable workhorses. Whatever else, they were a handsome lot, suggesting that the aesthetics of electric traction have many possibilities beyond those of a shoebox. And, since today is 'yesterday's tomorrow', it is worth remembering that the Baltics, now slightly antique in mien, were state of the art in their day.

When researching the story of No.13, many years ago, your author interviewed a long-retired Gateshead engine driver who had been one of the traction team attached to No.13. How did this Baltic electric strike observers when it first appeared for testing at Shildon? He pondered awhile, then stated (in broad Geordie to which standard English prose cannot do justice): 'Man, she wor like a spaceship'.

NOTES

1. The two acronyms refer to *Vereiniginge Deutsches Eisenbahn Verwaltungen*, Union of German Railway Administrations; and *Union Internationale des Chemins de Fer*.

2. Title of the author's book on the subject, published in 1970 - *The Electric Railway That Never Was, York-Newcastle 1919*.

3. See Erhard Born, *2C1, Entwicklung und Geschichte der Pazifik-Lokomotiven*, 1965, a product of the German school that likes to treat locomotives of any given wheel configuration, electric, steam or diesel, as members of a single category; electrically, Swiss and Bavarian examples predominate.

4. See, for example, R A S Hennessey, op cit; K Hoole, *The Electric Locomotives of the North Eastern Railway*, 1988; and K C Appleby, *Shildon-Newport in Retrospect*, 1990.

5. Raven was, not unlike a few other Chief Mechanical Engineers, broad-minded about forms of traction; he had well-developed ideas about diesel propulsion; his enthusiasm for electric traction was quite genuine: had not F W Webb of the LNWR forecast 100mph electric trains?

6. See *Locomotives of the LNER* Part 10b, RCTS, 1990

7. Loans from the Reconstruction Finance Corporation, and later the Public Works Administration, the latter being part of the 'New Deal' initiative. The full story of the many PRR electrification initiatives is told in M Bezilla, 'Electric Traction on the Pennsylvania Railroad 1895-1968', 1980; the saga of the P6 classes is described in Frederick Westing, Mike Bezilla and Roger L Keyser, *The Pennsy P5 Electrics*, 2002.

8. The ratio of horsepower to weight bearing on axles in these P9a machines was unmatched in PRR history, steam, diesel, or electric.

VERY HIGH VOLTAGE RAILWAYS

Sishen-Saldanha Railway, South Africa: seen here in July, 2009 - one of the Spoornet Class 9E, first series, Co-Co, 9016 of 1974; constructed by Union Carriage & Wagon Works (Transvaal) with GE electrical equipment. The high voltage of the line causes the insulators supporting the pantograph and busbars to be unusually high, set in the 'well' at the rear. Photo: *Colonel André Kritzinger*

The technical advantages of electric traction have been understood for well over a century. That said, the economic issues arising from electrification plans are also well known, principally its high capital cost.

A major source of this initial cost was the need to build substations along the line. Many early schemes, and virtually all tramway ones, were based on low voltage dc (500-600V mainly). High tension ac distribution was the norm, duly converted to low voltage in these substations. This was no easy matter before the development of obliging, unmanned rectifiers. Substations were generally large, even palatial, containing heavy rotary-converters (ac to dc) that had to be attended, sometimes by three shifts of 2-3 people each day.

The London, Brighton & South Coast Railway (LBSCR) scheme visited by SLS members in 1912[2] was different – a relatively high voltage ac system (6.6kV ac, 25 Hz) requiring far fewer substations than a dc system. Moreover, ac substations simply stepped down the voltage by means of transformers, an easier process than mechanical conversion.

Since that time the search has been on to gain further advantages from ac. For much of the first part of the last century, the 'grail' was to employ standard, industrial-frequency ac at 50Hz (60 Hz in the USA). The Hungarians attained it first (1932); by century's end it had become quite normal, as it is in the UK[3].

In practice the use of ac at 25kV is not quite as easy as this summary might imply. There are problems, for example, in maintaining the symmetry of single-phase supply from a national three-phase grid, on stretches of railway with widely varying loads. On the other hand, getting power supplies to a railway in a densely-settled industrial country like the UK, well covered by a National Grid, is a reasonably straightforward matter. However, in sparsely settled areas with limited power supplies, electrifying a railway has a further obstacle to overcome. One way round this can be the 'autotransformer' which feeds the catenary at intervals from a parallel supply of (say) 25kV at intervals of about 10km. But there is a further way of maintaining voltage over terrain where there is barely any human settlement: by raising the line voltage in the conductors to unusually high values, like 50kV. The outcome is the 'Very High Voltage' (VHV) railway, the subject of our enquiry.

This ploy has been used in a few isolated bulk-transport lines serving remote mining operations well away from major population centres. Remoteness is a crucial factor because using

59

the actual overhead conductor as a high voltage distributor in this way has considerable 'Health and Safety' issues, to put it mildly. Also, a very high voltage railway that had to thread a landscape well-endowed with road bridges would need much greater clearances than is normally the case. On the other hand, being isolated, these railways have no problems of systemic compatibility with other parts of the network.

Thus, whilst low voltage dc lines like the UK's south eastern network require substations at 2-3 mile intervals, the high voltage lines described here can extend intervals to 75 miles or more, in some cases even avoiding them altogether.

Perhaps because these railways were and are far away from major population centres they have not been very well known or recorded in mainstream literature, especially in the UK which has not had any such lines. Both tenses have to be employed because some of these very high voltage railways have come and gone well within living memory, as the summary following will demonstrate. Their remoteness has also contributed to their patchy coverage and the uncertain data surrounding some aspects of their record.

In 2000 there were five railways of this kind employing 'VHV' in their actual conductor wires, transforming the current aboard locomotives. Three remain, and there are some near misses worthy of note, lower voltage lines that are at once distant and yet prominent for one reason or another. The list below may not be complete; perhaps some readers can add to it.

Black Mesa & Lake Powell Rail Road (BMLPRR): a coal-carrying line in Arizona dating from the early 1970s, its conductors are energised at 50kV ac, 60Hz. It is 78 miles long and has one substation. Coal is no longer carried from Black Mesa mine, the original source, but from the nearby Kayenta mine. It is transported typically in trains of sixty 120-tonne wagons, '24/7' to the Navajo Generating Station, Page, Arizona, owned by the Navajo Tribal Utility Authority. Traction was originally six General Electric Company (GE) E60 locomotives, later supplemented by the improved E60-C type, bought in as-new condition from the unsuccessful Mexico City-Irapuato electrification (25kV ac), duly converted for 50kV operation, although paradoxically left in their 'National de Mexico' livery for a while.

The 80-mile Tumbler Ridge branch of the British Columbia Railway (BCR), dating from 1983, was another 50kV ac, 60Hz coal-carrier but rather 'greener' than the BMLPRR in that its power came from hydro-electricity. The loss of coal markets, notably in Japan, led to its de-electrification in 2000, and its folding in 2003, however plans were announced to reopen some of its operations in 2017[1].

Four locomotives of the Black Mesa & Lake Powell Rail Road, three (Class E60) in the line's own livery, one (Class E60C) still in National de Mexico hues. Photo: *Stu Levene*

The BCR became part of the Canadian National system shortly afterwards. The line was nearly the same length as the BMLPRR, but had two long tunnels Wolverine (3.6 miles) and Table (5.5 miles), hence the attraction of exhaust-free electric traction. Calculations at the time of its building suggested that setting up tunnel ventilation systems would have cost $Can 15 million. Furthermore, the line would have required 2 million gallons of diesel fuel per year as well as far costlier maintenance than electric traction.

In its operating years, the BCR employed seven leased General Motors (GM) Class GFC-6 Co-Co locomotives, hauling long trains of 91-tonne coal gondolas, so arranged with rotating couplers that they could be inverted to drop their load quickly; two at a time, 60 cars in an hour. These 6,000hp machines, which embodied some ASEA technology from Sweden, were scrapped when traffic dried up, but one was privately purchased and reposes in the British Columbia Railway and Forest Museum, Prince Rupert. Reports suggest that the catenary and conductors were left energised to discourage meddling and theft.

The Deseret Power Rail Road is a 35-mile line, 50kV ac, 60Hz, in the borderlands of Utah and Colorado, like the BMLPRR entirely isolated. Constructed in 1984 as the 'Deseret Western Railway', it has seven electric locomotives that run, generally, two 35-car trains daily connecting the Deserado coal mine to the Bonanza Power Plant. At each end of the line there is a loop, permitting in effect an 'MGO' fuel supply system. Motive power consists of two General Electric E60 freight electrics, with a further five E60C no longer required by the 'National de Mexico', as was the case with the Black Mesa line.

The Navajo Mine Rail Road is even remoter, if that were possible, 14 miles long, constructed in 1974 near Farmington, New Mexico. Trains usually have an electric locomotive at one end, a diesel-electric at the other, simply reversing roles as hauler and pusher at each end of the journey to and from the Four Corners Generating Station. Facts regarding this line are thin on the ground; it appears to have five electric locomotives in smart red, white and black livery, all from the GE E60 family, and is energised at either 25kV or 50kV ac; any further data gratefully received!

The Sishen-Saldanha Railway (SSR) in South Africa is the last of the VHV lines, still in robust operation. It is of 3ft 6in gauge, 535 miles long, constructed during 1973-76, and energised at 50kv ac, 50Hz.

Sometimes referred to as 'Orex' (Ore Export line) the SSR is dedicated to carrying iron ore in huge quantities from Sishen (4,249 ft above sea level) to the Atlantic at Saldanha Bay, Western Cape. The line is single, with nineteen passing loops. Typical trains have been composed of 210 ore wagons of 91 tonnes each (similar to the BCR, 91 tonnes being

Sishen-Sandalha Railway, two of the latest 50kV locomotives by Union Carriage & Wagon Works and Mitsui, Class 15E, No.15 006 and 15 001 at Salkor Yard, Sandalha, 19 August 2010. Photo: *Colonel André Kritzinger*

100 short tons, the original specification). Some record loads involving ten locomotives (electric and diesel electric) have been listed; ore-caravans of 342 wagons, the spaced-out traction being radio-controlled. So long are these trains that crews are given motor scooters, stashed away on locomotives, to get around the train on regular tours of inspection.

Being longer than its North American cousins, the SSR has six substations, albeit with a noticeable voltage drop on some sections causing motive power to be designed to cope with a range of 25-55kV. Locomotives are Co-Co types of Class 9E (locally constructed, with electrical equipment by GE UK), now being supplemented by thirty-two Class 15E, also of Co-Co configuration.

All these lines have, or had, exceptionally high productivity because little labour was required to shift immense tonnages. None actually achieved driverless trains, although the BMLPRR experimented along these lines.

One mineral line that did reach this goal for a while was the **Muskingum Electric Rail Road** (MERR) in Ohio, USA, although it operated on the more modest and familiar 25kV ac, 60Hz, high, however, by the standards of traction in the Americas where it was the first high voltage line. It is now defunct following mine closures. This 20-mile line was isolated from other railways, except for a single interchange with the CSX system for the supply of equipment. The driverless, automated locomotives were General Electric Class E50C, two of which, with about 100 hoppers or 'gons' (USA – gondolas) kept up a conveyer-belt service, some 15 wagons per train. As a train arrived at its loading destination it would automatically slow to about 1/3 mph, creeping by the tipple. Trains were in fixed formations, moving to and from each terminus without recourse to loops.

At the power station terminus of the MERR train control was carried out by shoes on the hoppers which picked up signals and opened their doors. At regular intervals along the track further command signals kept the robot train under firm control. The line lasted 1968-c2002, closing when the strip mining was but 2 miles from the power station. Interestingly, the MERR ran two private, office cars (*Dover Fort* and *Oak Lane*) – one wonders if these, too, were hauled about robotically?

In Germany, the **Rübelandbahn**, is an outpost of relatively high tension 25kV ac 50Hz working in a sea of the central and Northern Europe railway standard of 15kV ac, 16.6Hz. It, too, is another mineral line, carrying limestone in the Harz mountains.

First electrified by the DR in 1966, the equipment became run down and was replaced by diesels over the years. DR's successor, DB Netz was disinclined to refurbish the line once the limestone operators (Fels Werke) took a lease on the line through a subsidiary, Fels Netz, and then awarded its transport contract to a private operator, HVLE (Havelländische Eisenbahn) – originally a line operating in the Berlin area, but now spreading its wings. Although technically 'private', HVLE has large public authority shareholdings.

During the first phase of electrical operation, the Rübelandbahn used fifteen DB series 251 Co-Co locomotives. Since the heavy refurbishment carried out by the new regime (2006-08), HVLE has purchased two Bombardier TRAXX (**T**ransnational **R**ailway **A**pplications with e**X**treme fle**X**ibility) Bo-Bo locomotives, impressively decked out in the company's grey and orange livery. The upgraded and reborn ac system came on stream in April, 2009.

Because the VHV lines have all been located in fairly remote places, reliant upon isolated power systems and not much concerned with the clearance problems, they are, perhaps, good examples of the 'courses for horses' doctrine, and would not therefore supply a basis for the kind of networks required in densely settled Europe. The notion of VHV driverless trains, London-Glasgow, could only bring on Health and Safety swoons.

The commonest motive power on the VHVs has been the GE E60 type, some of them supplied in virtually new condition from the National Railways of Mexico, or rather its privatised successor. Some came from US sources, like the Navajo Mine Rail Road's two from Amtrak and one from the New Jersey Transit Authority. Strictly speaking the E60 has various sub-classes; rostermongers have not always been assiduous in stating which applies to the VHV lines (or they disagree). Even so, the characteristics of the various clan members are much the same.

The E60s were intended to offer an American standard electric Co-Co (C-C in American usage) of great power. They were constructed in two batches (E60 and E60C), 1972-83, embodying thyristor control which extended to braking and wheelslip regulation. The E60C-2 type acquired from Mexico had cabs at each end whereas the first batch sent to the Black Mesa line only had single cabs. The outcome either way was a beefy-looking 6,000hp machine, not unattractive, able to attain over 70mph, higher at first, although some derailments led to the fixing of a more modest maximum. By way of increasing adhesion, the BMLPRR versions bear 45 tonnes of ballast (total weight 193 tonnes). Their six GE780 motors (producing 77,000 lb tractive effort), were mounted in the time-hallowed nose-suspended and axle hung form, straightforward if none too kind on the permanent way.

The Tumbler Ridge fleet of GF6-C locomotives, all single-cab Co-Cos, was constructed by General Motors, Electro-Motive Division (more usually associated with diesel-electrics) using Swedish equipment from ASEA, including thyristors and radar-operated wheelslip detectors. Rheostatic braking involved large, roof-mounted resistances cooled by fans.

The twenty-five Sishen-Sandalha locomotives, Class 9E were entirely different machines, South African built with British electrical equipment – in some respects the last hurrah of GE Traction and,

Three E60 electrics hauling a coal train through wintry terrain on the Deseret Power Rail Road, near the Bonanza power station; two E60s in front, leading a former National de Mexico E60C still in its original livery. Photo: *Dick Ebright*

in their day, the world's most powerful 3ft 6in gauge locomotives. They had air-conditioned cabs at one end only, a tractive effort of 121,000lb and a top speed of 56mph. Their successors, seventy-six Class 15E, are Co-Co electrics of 6,000hp first delivered in 2010, part of an order due to last until 2013. Although still constructed in South Africa, the 15Es arrived long after GE Traction had gone down the tubes. Design is Japanese, from Mitsui with Toshiba electrics.

There may be other candidates for the VHV laurels, and it would be interesting to learn about them.

NOTES

1. See news item at http://www.cbc.ca/news/canada/british-columbia/cn-agrees-to-reopen-tumbler-ridge-rail-line-to-start-shipping-coal-1.3969825

2. As mentioned in the Foreword one of the first outings by members of what is now The Stephenson Locomotive Society was to view the, then new, London, Brighton & South Coast Railway electric car sheds near Norwood in South London.

3. Apart, that is from the 750V dc lines in London and South East England, tramways, and metro systems (London, Glasgow) all of which adhere to this historically-determined low voltage arrangement. Recent (June 2011) statements by Network Rail suggest that the famous third rail system of the former 'Southern Electric' may be unsustainable under modern conditions and levels of demand, although replacing it with overhead conductors will be neither easy or cheap.

REFERENCES

As noted, records of some of these lines are sparse. See: John B Corns, 'Ohio's Robot Railway', *Trains Magazine*, March 1979; and, briefly on the BM&LP, *Trains*, October 1974. See also Y Machefert, F Nouvion & J Woimant, *Histoire de la Traction Electrique*, 1986; J Edward Martin, *On a Streak of Lightning, Electric Railways in Canada*, 1994 and, of course, the Web.

The main part of this chapter giving details of individual railways first appeared, in shorter form, in the Electric Railway Society Journal; it is reproduced here by kind permission of its Editor, John Rosser.